15

REEDS MARINE ENGINEERING AND TECHNOLOGY

ELECTRONICS, NAVIGATIONAL AIDS AND RADIO THEORY
FOR ELECTROTECHNICAL OFFICERS

Steve Richards

REEDS
LONDON • OXFORD • NEW YORK • NEW DELHI • SYDNEY

REEDS
Bloomsbury Publishing Plc
50 Bedford Square, London, WC1B 3DP, UK
29 Earlsfort Terrace, Dublin 2, Ireland

BLOOMSBURY, REEDS, and the Reeds logo are trademarks of Bloomsbury Publishing Plc

First published in Great Britain 2013
Reissued 2019
This edition published 2023

A catalogue record for this book is available from the British Library

Library of Congress Cataloguing-in-Publication data has been applied for.

ISBN: PB: 978-1-3994-1002-1; ePub: 978-1-3994-1003-8; ePDF: 978-1-3994-1004-5

2 4 6 8 10 9 7 5 3 1

Typeset in Myriad Pro 10/14 by NewGen
Printed and bound in Great Britain by CPI group (UK) Ltd, Croydon CR0 4YY

To find out more about our authors and books visit
www.bloomsbury.com and sign up for our newsletters

REEDS MARINE ENGINEERING AND TECHNOLOGY

ELECTRONICS, NAVIGATIONAL AIDS AND RADIO THEORY

FOR ELECTROTECHNICAL OFFICERS

REEDS MARINE ENGINEERING AND TECHNOLOGY SERIES

Vol. 1 Mathematics for Marine Engineers

Vol. 2 Applied Mechanics for Marine Engineers

Vol. 3 Applied Heat for Marine Engineers

Vol. 4 Naval Architecture for Marine Engineers

Vol. 5 Ship Construction for Marine Students

Vol. 6 Basic Electrotechnology for Marine Engineers

Vol. 7 Advanced Electrotechnology for Marine Engineers

Vol. 8 General Engineering Knowledge for Marine Engineers

Vol. 9 Steam Engineering Knowledge for Marine Engineers

Vol. 10 Instrumentation and Control Systems

Vol. 11 Engineering Drawings for Marine Engineers

Vol. 12 Motor Engineering Knowledge for Marine Engineers

Vol. 13 Ship Stability, Resistance and Powering

Vol. 14 Stealth Warship Technology

Vol. 15 Electronics, Navigational Aids and Radio Theory for Electrotechnical Officers

Vol. 16 Electrical Power Systems for Marine Engineers

CONTENTS

PREFACE

The role of the ETO, post STCW 2010 (Manila Amendments), has become recognised internationally. Ships have increasing amounts of electronic equipment on board, driving the need for shipping companies to recruit dedicated ETOs. This book addresses that part of the ETO function that deals with the electronic equipment located on or near the ship's bridge.

The Association of Marine Electronic and Radio Colleges (AMERC) sets the syllabus for ETO examinations in: Electronic Principles; Electronic Navigation Electrical Maintenance Certificate (ENEM); and the GMDSS Radio Maintenance Certificate. This book specifically covers the material needed to assist in the understanding of the technology used in these examinations.

It addresses each topic in a simple and matter of fact way that will be easy for the student to understand. I wrote this book to help the ETO student with their studies and for their first few years at sea.

I would like to thank all the companies who supplied images of their marine equipment for use in this work.

1 BASIC ELECTRONICS

Insulators and Conductors

All matter is made up of atoms. Atoms are made up of electrons, protons and neutrons. The protons and neutrons exist together in a tight ball called the nucleus surrounded by orbiting electrons. The electrons are negatively charged, the protons within the central core of the atom are positively charged. Different elements contain different amounts of protons, neutrons and electrons. Some atoms have loose electrons in their outer shell; other elements have electrons tightly held within each orbit. Atoms in copper (Cu) have very loose electrons in its outer shell, atoms on glass (mainly SiO_2) have very tightly bound electrons. Loose electrons can be made to move through an element if the element is subject to an electrical force. Tightly bound electrons do not move, even when subject to a large electrical force (Figure 1.1).

Copper allows the easy movement of electrons. Copper is called a conductor because it allows the easy movement of electrons. Glass does not allow the movement of electrons. Glass is called an insulator because it does not allow the easy movement of electrons. Table 1.1 identifies some popular conductors and insulators.

The larger the cross-sectional area (CSA) of a conductor, the more electrons are available to move from one end of the conductor to the other. The smaller the CSA of a conductor, the smaller the amount of electrons available to move from one end of the conductor to the other. The opposition that elements give to the movement

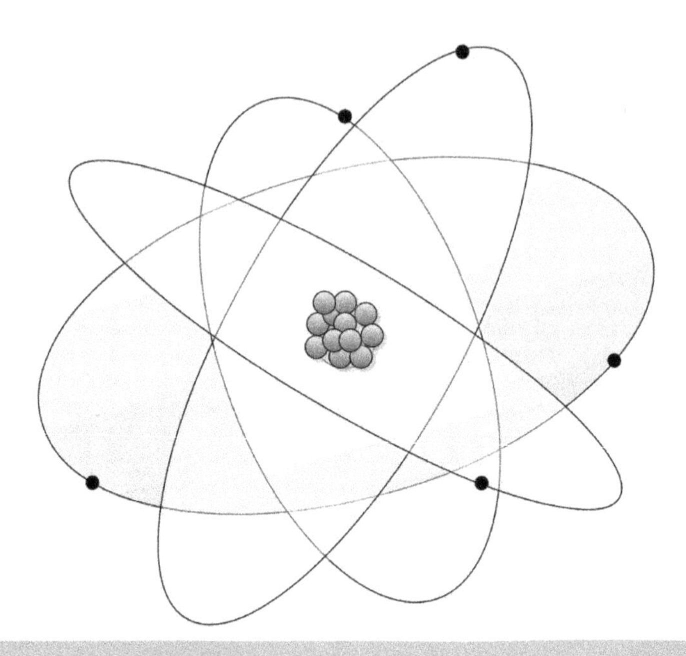

▲ **Figure 1.1** *Layout of an atom*

Table 1.1 *Typical conductors and insulators*

Conductors	Insulators
Silver	Glass
Copper	Rubber
Gold	Porcelain
Aluminium	Ceramic
Brass	Plastic
Bronze	Air
Mercury	Pure water
Graphite	Quartz

of electrons when under the influence of voltage is called resistance. Electrons, being negatively charged particles, move away from a negative force or potential, and are attracted to a positive force or potential. The number of electrons that move depends upon the force (voltage) and the opposition (resistance) offered by elements from which the conductor is made.

The movement of electrons in a conductor is called an electrical current and is measured in Amperes (A). The opposition to electron flow offered by an element is called its resistance, and is measured in ohms (Ω). The higher the value in ohms, the greater the resistance to electron flow.

The amount of electron flow is proportional to the force (voltage) applied to the element offering resistance, and inversely proportional to the amount of resistance. Ohm developed the equation given his name:

Ohm's law

$$I = \frac{V}{R}$$

where I is the current in Amperes, V is the voltage pressure or force applied measured in Volts and R is the resistance measured in Ohms. Since Ohm's law is a simple equation, it can be transposed to find: V, R or I.

$$V = IR, \quad R = \frac{V}{I}, \quad I = \frac{V}{R}$$

Resistance

Ohm's law tells us how much current (electron flow) there will be if we apply a voltage to a wire which has a resistance value. Sometimes we need good conductors; other times we need very bad conductors. Examples of each type of resistance value show that we need different values for resistance: Copper is a good conductor, plastic is a bad conductor, the two form a perfect pair when they are assembled into an insulated wire. The copper provides a path for electrons to flow through due to its low resistance, while the plastic surrounding the copper prevents any electrons from 'leaking' away. Copper wire has a very low resistance, depending upon its CSA. Most plastic has a high resistance depending upon the elements used in its manufacture and its thickness. There are occasions when we need a resistance value that falls between the very low (copper) and very high (plastic). We can obtain devices which possess a specified amount of resistance, we call these devices resistors. Resistors can be obtained in a variety of values, for example:

1R0	10R	100R	1K0	10K	100K	1M0
1R2	12R	120R	1K2	12K	120K	1M2
1R5	15R	150R	1K5	15K	150K	1M5
1R8	18R	180R	1K8	18K	180K	1M8
2R2	22R	220R	2K2	22K	220K	2M2
2R7	27R	270R	2K7	27K	270K	2M7
3R3	33R	330R	3K3	33K	330K	3M3
3R9	39R	390R	3K9	39K	390K	3M9
4R7	47R	470R	4K7	47K	470K	4M7
5R6	56R	560R	5K6	56K	560K	5M6
6R8	68R	680R	6K8	68K	680K	6M8
8R2	82R	820R	8K2	82K	820K	8M2

The above table illustrates that not every resistance value can be purchased. Due to the small size of resistors, a colour coding scheme is used to communicate the resistance value. Four or five bands of different colour identify each resistor value. The first band is the first digit of the value, the second band is the second digit of the value, the third band is the multiplier (number of zeros), the fourth band the tolerance value. For a resistor whose coloured bands are red, red, yellow and gold the values are 2, 2, 0000, 5%, which is 220 kΩ 5% (Table 1.2).

Table 1.2 *Resistor colour code*

Number	Colour
0	Black
1	Brown
2	Red
3	Orange
4	Yellow
5	Green
6	Blue
7	Violet
8	Grey
9	White

On some large resistors, there is space to write the resistance value on the body of the resistor. An abbreviated short hand is used to save space: The decimal point is changed into an R, K, M or G signifying the power of 10 in use. See the BS 1852 coding examples below:

R33	0.33 Ω
2R2	2.2 Ω
470R	470 Ω
1K2	1.2 kΩ
22K	22 kΩ
22K2	22.2 kΩ
4M7	4.7 MΩ
5K6G	5.6 kΩ 2%
33KK	33 kΩ 10%
47K3F	47.3 kΩ 1%

Resistors are used throughout electronics for a variety of reasons, which we will discuss next. The electronic circuit symbol for a resistor is shown in Figure 1.2.

▲ **Figure 1.2** *Symbol for a resistor*

When a flow of electrons is pushed through a resistor, the electrons use up energy moving through the resistor. We can connect a source of electrons (a battery) to the resistor, such that electrons can flow out of the battery, through the resistor returning to the battery. We can see this circuit in Figure 1.3.

It can be seen that electrons flow out of the negative terminal of the battery, through the resistor, then into the positive terminal of the battery.

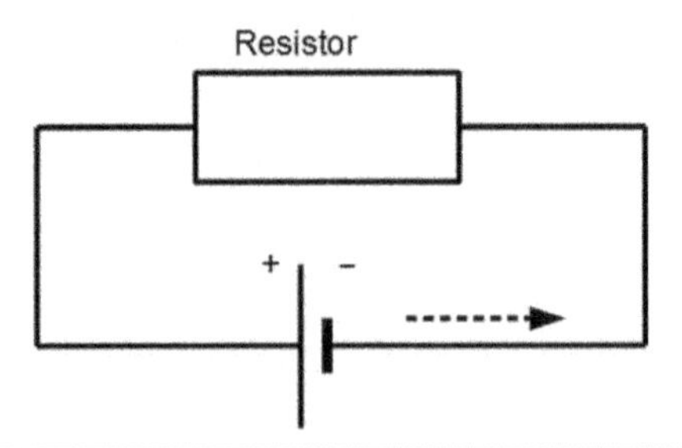

▲ **Figure 1.3** *Simple circuit with resistor and battery*

Series and parallel resistors

As resistors are not available in all desired values, we have to connect resistors together to reach our required value. Resistors can be connected together in two different ways: series or parallel.

Resistors can be connected together in series.

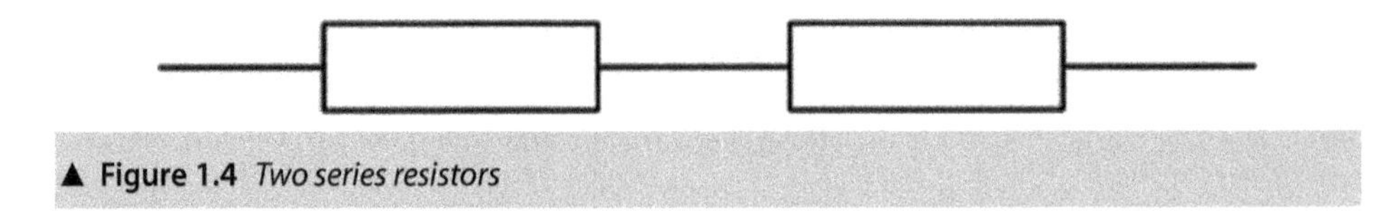

▲ **Figure 1.4** *Two series resistors*

In Figure 1.4, the electrons have to flow through one resistor then the other resistor. In effect, two resistors in series have the same value as one resistor of the same combined value. As shown in Figure 1.5, the two 100 Ω resistors are equal to one 200 Ω resistor. To calculate the total value of resistors connected in series, simply add the individual values together.

For example, a 500 Ω resistor in series with a 200 Ω resistor is 500 + 200 = 700 Ω.

When connecting resistors in parallel, we need to use a different method to calculate the total resistance (Figure 1.6).

We calculate the combined value by using one of two methods, product over sum method or the reciprocal method.

Product over sum.

$$R_{\text{Total}} = \frac{\text{R1} \times \text{R2}}{\text{R1} + \text{R2}}$$

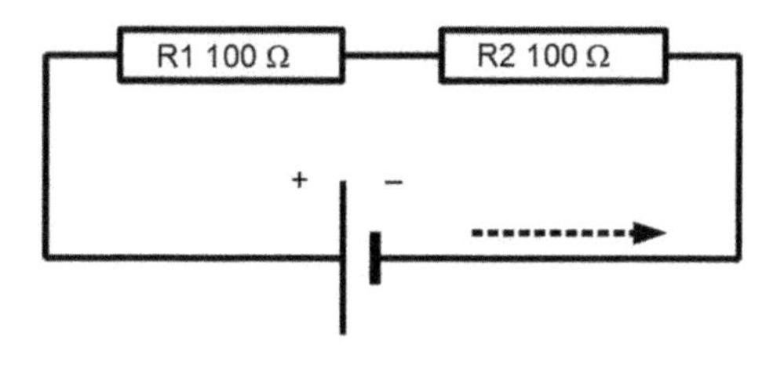

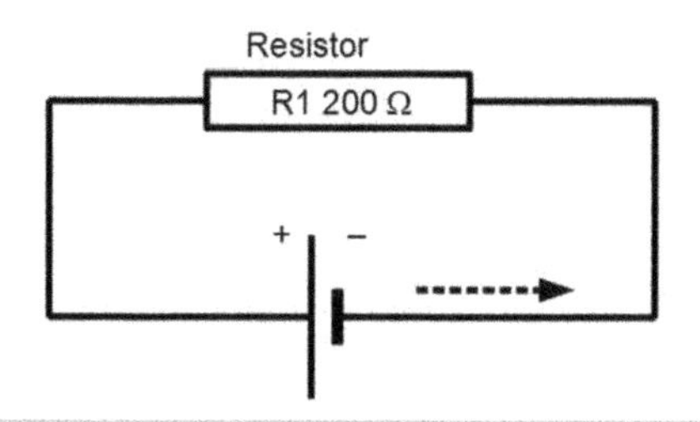

▲ **Figure 1.5** *Two series resistors, same as one resistor of combined value*

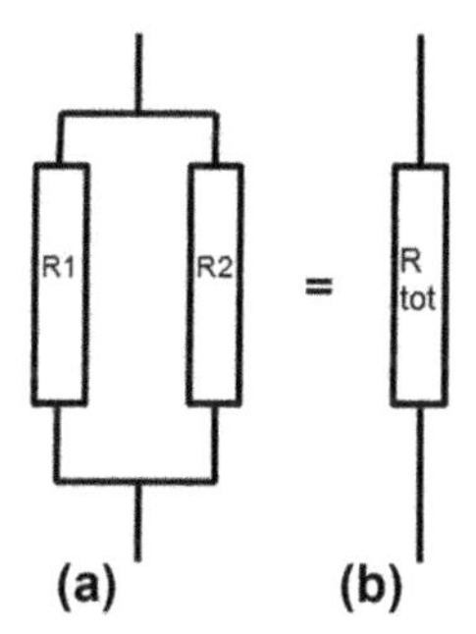

▲ **Figure 1.6** *Two parallel resistors, same as one resistor of combined value*

This method can be used when there are two resistors in parallel. The answer in ohms will always be lower than the lowest resistor value,

for example, R1 = 200 Ω, R2 = 300 Ω.

$$R_{\text{Total}} = \frac{200 \times 300}{200 + 300} = \underline{120\,\Omega}$$

If you have three or more resistors in parallel then the second method can be used.

$$R_{\text{Total}} = \frac{1}{\frac{1}{\text{R1}} + \frac{1}{\text{R2}} + \frac{1}{\text{R3}}}$$

This is more usually written as:

$$\frac{1}{R_{\text{Total}}} = \frac{1}{\text{R1}} + \frac{1}{\text{R2}} + \frac{1}{\text{R3}}$$

for example,

$$\text{R1} = 100\ \Omega,\ \text{R2} = 200\ \Omega,\ \text{R3} = 75\ \Omega$$

$$\frac{1}{R_{\text{Total}}} = 0.0283$$

$$R_{\text{Total}} = \frac{1}{0.0283} = \underline{35.29\ \Omega}$$

In Figure 1.7(a) two parallel resistors (R1 and R3) are in series with R2. The total value of resistance of this circuit can be visualised as Figure 1.7(b) where the two parallel resistors have been substituted by one resistor of equivalent value. R1 in parallel with R3 is 50 Ω. Figure 1.7(b) can be visualised as two resistors in series, the total value of which is

$$100\ \Omega + 50\ \Omega = 150\ \Omega$$

We can say that Figure 1.7(c) is the equivalent circuit of Figure 1.7(a). A measuring instrument – digital volt meter (DVM) or digital multi-meter (DMM) – would not be able to tell the difference between (a), (b) or (c), they would all measure 150 Ω.

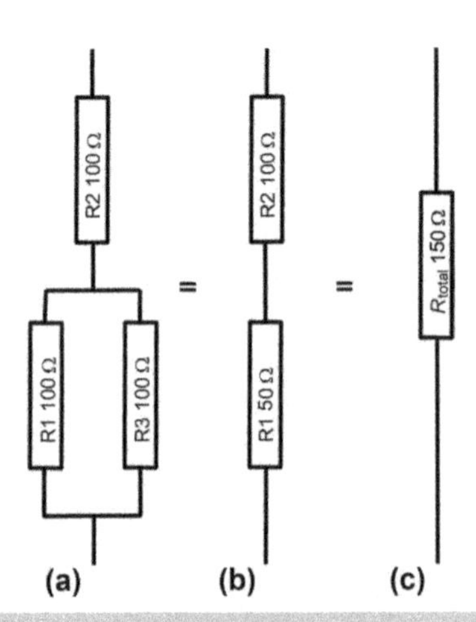

▲ **Figure 1.7** *Two parallel resistors in series with one resistor*

In Figure 1.8, calculate the voltages at point A and B with respect to point C, and the current flowing through R3.

First, calculate the total circuit resistance.

$$R_{\text{Total}} = \text{R1} + \frac{\text{R2} \times \text{R3}}{\text{R2} + \text{R3}} + \text{R4}$$

$$R_{\text{Total}} = 100 + \frac{100 \times 100}{100 + 100} + 50$$

$$R_{\text{Total}} = \underline{200\Omega}$$

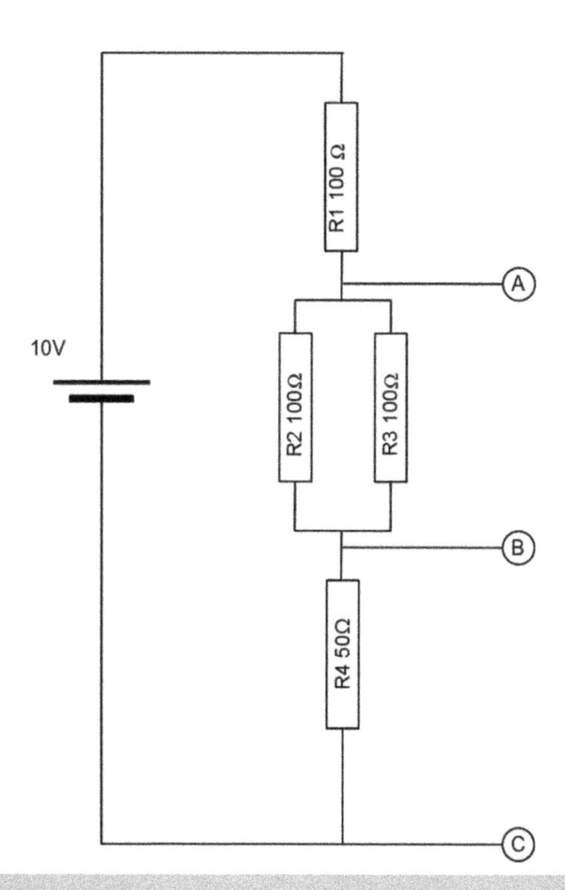

▲ **Figure 1.8** *Example 1 – resistors*

Now calculate the total circuit current:

$$I_{\text{Total}} = \frac{V}{R} \text{ Amps}$$

$$I_{\text{Total}} = \frac{10}{200} \text{ Amps}$$

$$I_{\text{Total}} = \underline{0.05\text{ A}} \text{ or } \underline{50\text{mA}}$$

Now we have the circuit current, which is the current passing through R1 and R4, we can calculate the voltage across the pair of parallel resistors R2 and R3.

To calculate the voltage across the parallel pair R2 and R3 we need to calculate the voltage dropped across both R1 and R4, and the remaining voltage from the 10 V supply will be the voltage across R2//R3.

$$V \text{ across } R_1 = I \times R \text{ Volts}$$

$$V_{\text{R1}} = 0.05 \times 100$$

$$V_{\text{R1}} = 5\text{V}$$

$$V_{\text{R4}} = 0.05 \times 50$$

$$V_{\text{R4}} = 2.5\text{ V}$$

The voltages consumed by this circuit so far are: V_{R1}=5 V and V_{R4}=2.5 V, a total of 7.5 V.

This voltage when subtracted from the supply voltage: $10\text{V} - 7.5\text{V} = 2.5\text{V}$ equals the voltage across the parallel pair R2//R3 ($V_{\text{R2//R3}}$).

We now know all the voltages in this circuit. Voltage at point B with respect to point C is:

$$V_{\text{R4}} = \underline{2.5\text{V}}$$

Voltage at point A with respect to point C is:

$$V_{\text{R4}} + V_{\text{R2//R3}} = 2.5\text{V} + 2.5\text{V} = \underline{5\text{V}}$$

The current through R3 is:

$$I_{R3} = \frac{V_{R2//R3}}{R3}$$

$$I_{R3} = \frac{2.5}{100} A$$

$$I_{R3} = \underline{0.025\ A} \quad \text{or} \quad \underline{25\,mA}$$

Resistors dissipate heat when an electric current flows through them. The amount of heat generated can be calculated by three related formulas:

$$P = VI \text{ Watts}, \quad P = I^2R \text{ Watts}, \quad P = \frac{V^2}{R} \text{ Watts}$$

The amount of power that a resistor can dissipate without failing depends upon its size and construction. A larger resistor has a larger surface area enabling heat to dissipate more readily. Resistors can be made from high temperature materials allowing them to dissipate more heat without failing. Resistors are made in a range of standard wattages:

1/8 W, 1/4 W, 1/2 W, 1 W, 2 W, 3 W, 5 W, 10 W, 25 W, etc.

Higher power resistors (above 2 W) usually need to be bolted to a metal heat sink to enable the dissipation of heat. A resistor can carry the maximum designated current without failing when installed correctly.

Questions

In Figure 1.9, calculate the following:

1. Voltage at point A with respect to point C (2.525 V)
2. Voltage at point B with respect to point C (1.895 V)
3. Current flowing through R3 (25.25 mA)
4. Current flowing through R4 (37.9 mA)
5. Total power dissipated in the circuit (454.8 mW)
6. Minimum wattage for R3 (1/8 W or 1/4 W)

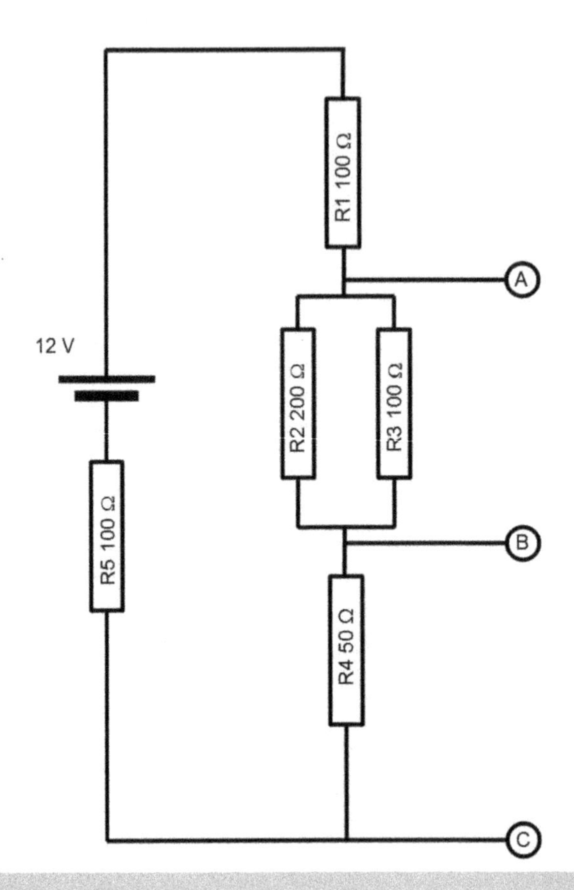

▲ **Figure 1.9** *Example 2 – resistors*

In Figure 1.10, calculate the following:

1. Voltage difference between points A and B (3 V)
2. Current flowing through R3 (30 mA)
3. Current flowing through R4 (45 mA)
4. Total power dissipated in the circuit (540 mW)
5. Suitable wattage of R4 (1/8 W or 1/4 W)

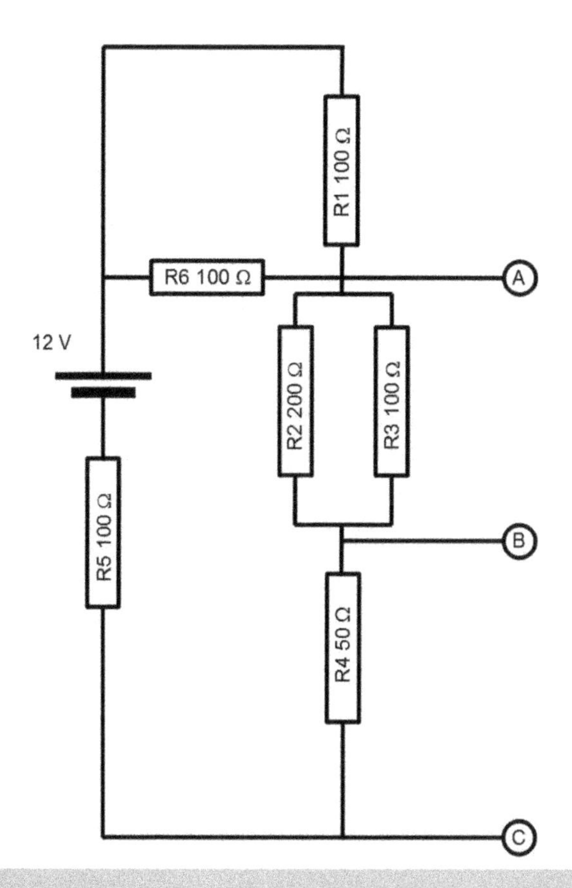

▲ **Figure 1.10** *Example 3 – resistors*

Capacitance

Capacitance is a property that exists when two conductors are separated by an insulator. A capacitor can store electrical energy. When a charged body is brought close to another charged body, the bodies either attract or repel one another. (If the charges are alike they repel; if the charges are opposite they attract.) The field that causes this effect is called the electrostatic field. The amount by which two charges attract or repel each other depends upon the size of the charges and the distance between the charges. The electrostatic field (force between two charged bodies) may be represented by lines of force drawn perpendicular to the charged surfaces. The current view is that the voltage pressure difference between the two bodies causes electrons to have a deformed orbit around the nucleus of each atom. The higher the voltage, the greater the deformation. The effect of deformation is realised as an electrostatic field. Figure 1.11 shows a two plate capacitor connected to an electrical cell. Electrons flow through the conductors to each of the metal plates. As the electron orbits are deformed, electrons flow into the negative plate and flow out of the positive plate. The energy stored in the deformed orbits manifests as an electrostatic field between the two plates.

The amount of energy that a capacitor can store is determined by its physical construction. A capacitor is usually constructed with two plates of conductive material separated by an insulator. The closer the two plates of metal, the greater the amount of energy that the capacitor can store, hence, the greater the value of the capacitor. The larger the area of the two plates of metal, the greater the value of the capacitor. The type of material used for the insulator affects the value of capacitance and hence the amount of energy that can be stored in the capacitor. A capacitor's ability to store energy is measured in Farads (F). The Farad is a very large unit of measurement. In most electronic systems, typical values of capacitors are micro Farads (μF) = 10^{-6} and pico Farads (pF) = 10^{-12}.

In Figure 1.12 we see a battery connected across a capacitor via a switch. In Figure 1.12(a) we see that the capacitor is empty, not charged, no voltage can be measured across the plates of the capacitor. In Figure 1.12, we operate the switch to connect the battery to the capacitor. Immediately, electrons start to flow out of the negative side of the battery

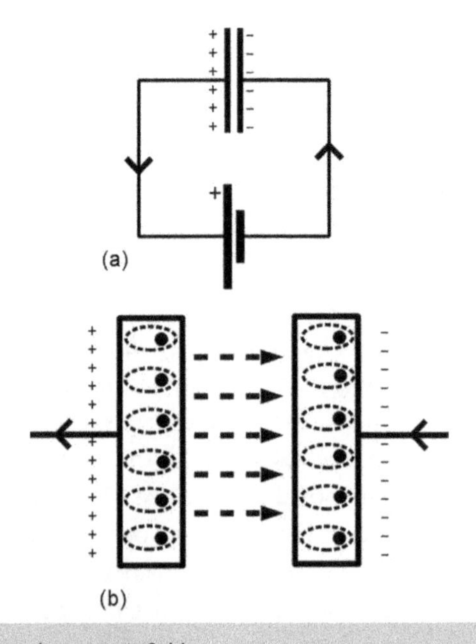

▲ **Figure 1.11** *Capacitor electrostatic field*

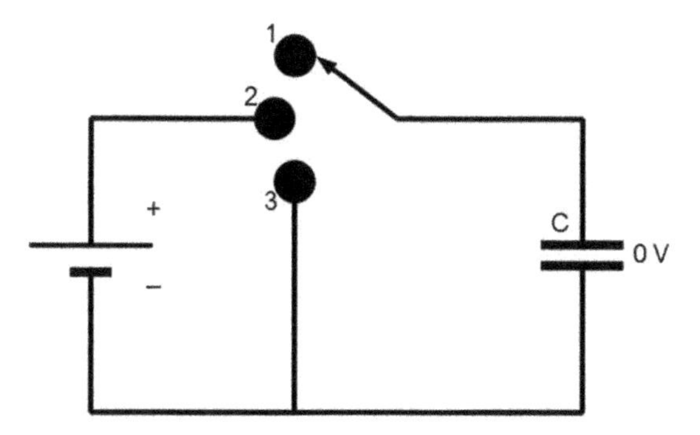

(a) Switch open, capacitor not charged

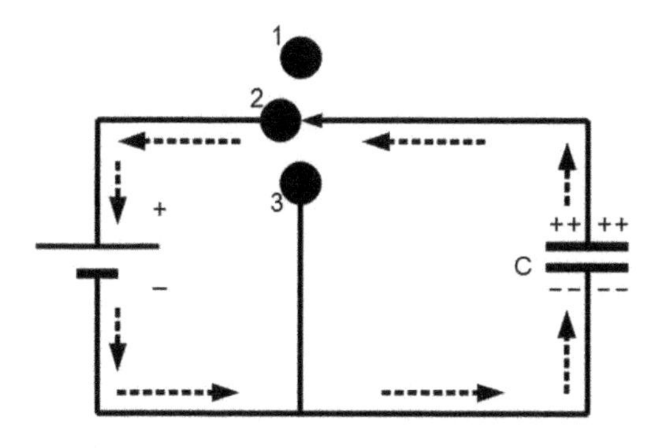

(b) Switch closed, capacitor charged

▲ **Figure 1.12** *Capacitor charging*

into the capacitor. Simultaneously, electrons are pulled out of the other side of the capacitor, into the positive side of the battery. For each electron that goes into the top of the capacitor and each electron that is dragged out of the lower half of the capacitor, a small voltage develops across the capacitor. Each electron movement increases this voltage. Each electron movement deforms electron orbits. When sufficient electrons have entered the top plate of the capacitor, and left the bottom plate of the capacitor, the voltage created by these electrons will be equal to the battery voltage (10 V). When the voltages are equal (battery voltage = capacitor voltage) then the reason for electrons to flow has disappeared. In Figure 1.13 we have moved the switch to break the connection between the battery and capacitor. The capacitor remains charged, retaining excess electrons on the negative plate, and a shortage of electrons on its positive plate. The capacitor will remain charged while its electrons do not leak away. The amount of electron loss or leakage depends upon the quality of the insulator between the two plates of the capacitor. In Figure 1.13(b) we have moved the switch to apply a short circuit across the capacitor. Electrons from the negative plate will now

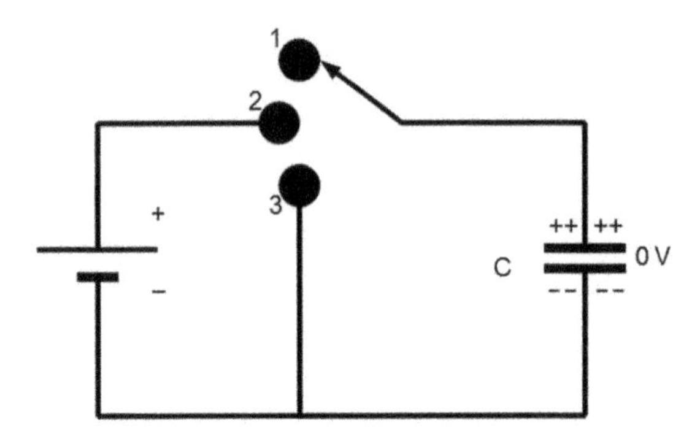

(a) Switch open, capacitor retains charge

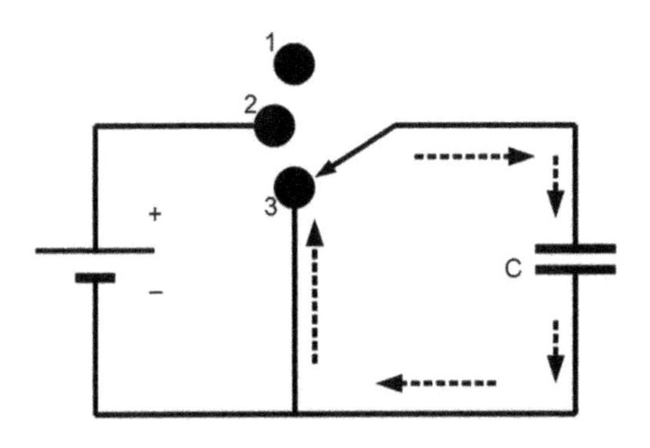

(b) Switch closed, capacitor discharged

▲ **Figure 1.13** *Capacitor discharging*

rush through the switch to the positive plate, where a shortage of electrons (positive charge) awaits them. As electrons leave the capacitor, the voltage across the capacitor decreases. When the last additional electron has left the plates of the capacitor, the voltage across the plates is zero.

In reality, it is unwise to connect a battery directly across a large capacitor, or to short out a large capacitor. This is due to the very high current (surge of electrons) that will flow into or out of the capacitor. Many circuits that involve capacitors use a resistor in series with the capacitor to limit the flow of electrons (current) into or out of a capacitor.

In Figure 1.14 we have a battery connected to a series capacitor and resistor via a switch. The instant we close the switch, the full battery voltage will appear across the series resistor/capacitor pair. Current will start to flow through the resistor, enabling the capacitor to charge up toward the battery voltage. As the capacitor fills up with charge,

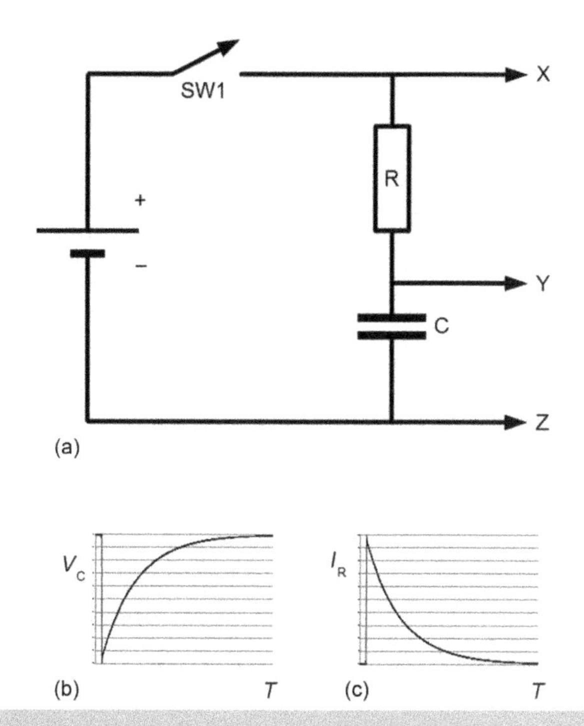

▲ **Figure 1.14** *Capacitor charging through resistor*

the voltage across it increases, therefore, the voltage across the resistor decreases. Eventually the capacitor reaches the battery voltage, current stops flowing and the capacitor has reached maximum charge. The shape that the capacitor voltage follows is exponential. It is notable that if you multiply the value of the capacitor (in Farads) with the value of the resistor (in Ohms), you obtain a value in seconds that is equal to 63.2% of the maximum voltage across the capacitor. This time is called the CR time constant. As the voltage rise across the capacitor follows an exponential curve, the maximum voltage possible across the capacitor is never reached. It is convention to assume that after 5CR, the capacitor has, for all practical purposes, reached its maximum voltage. Discharging the capacitor through the resistor produces a very similar exponential curve.

In Figure 1.15 we can see a complete charge discharge curve of the voltage across the capacitor. Both charge and discharge follow the distinctive exponential curve. It is important to note that due to the small distance between the plates of typical capacitors, an upper voltage limit exists. If this limit is exceeded, the electrical stress across the two plates is sufficient to allow an electrical arc to jump from one plate to the other. Depending upon the material between the two plates, a permanent short circuit can be formed, causing potential equipment damage or fire risk.

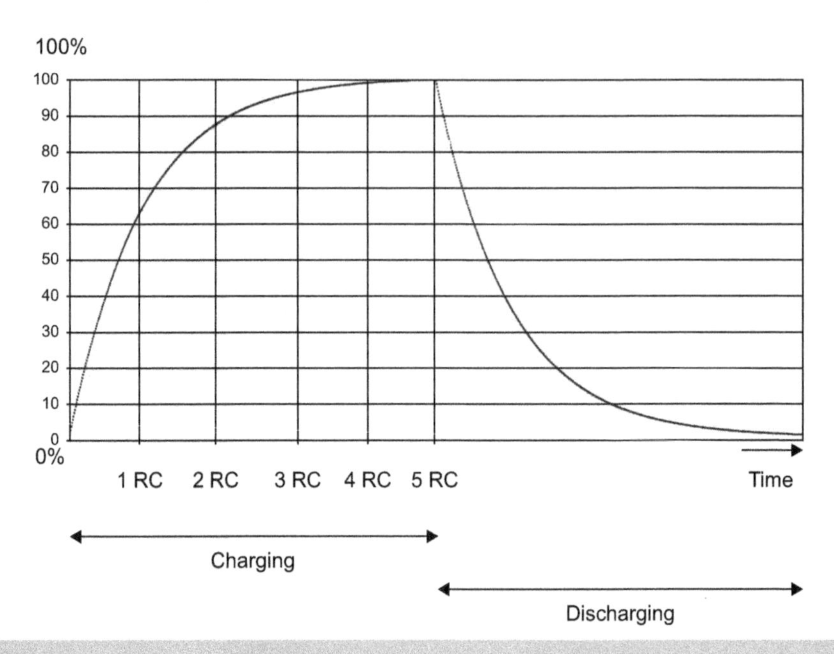

▲ **Figure 1.15** *Capacitor charge discharge cycle*

Series and parallel capacitors

Similar to resistors, capacitors are available in a limited range of values. Sometimes it is necessary to connect capacitors together in different combinations to reach a required value. Connecting two capacitors in parallel is the same as having one capacitor whose surface area is the addition of the two smaller surface areas, for example, to find the total effective value of two or more capacitors in parallel, add the values together (Figure 1.16).

If you have two capacitors in series, the effective total capacitance is calculated by using either product over sum or the reciprocal method.

That is,

$$C_{\text{Total}} = \frac{\text{C1} \times \text{C2}}{\text{C1} + \text{C2}}\,\text{F}$$

or

$$C_{\text{Total}} = \frac{1}{\frac{1}{\text{C1}} + \frac{1}{\text{C2}} + \frac{1}{\text{C3}}}$$

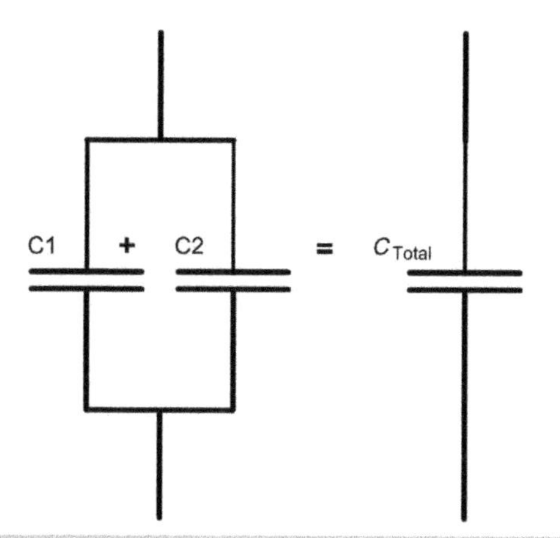

▲ **Figure 1.16** *Capacitors in parallel*

This is more usually written as:

$$\frac{1}{C_{Total}} = \frac{1}{C1} + \frac{1}{C2} + \frac{1}{C3} + \cdots$$

Of course, the second method can be used for an unlimited number of capacitors in series (Figure 1.17).

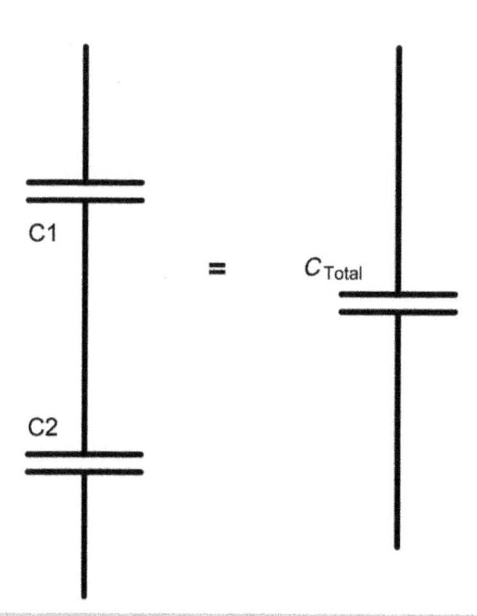

▲ **Figure 1.17** *Capacitors in series*

When we connect capacitors in series, the charge (the number of electrons) is the same for each capacitor in the circuit. (Being a series circuit, electrons travelling from the battery negative terminal travel through each capacitor before returning to the battery positive terminal.) As the number of electrons (or charge) is the same at any point in a series circuit, then the charge in each capacitor must be the same, and the same value for the whole circuit.

Charge (Q) measured in coulombs $= CV$ where C is the capacitance in Farads and V is the voltage across the capacitor.

$$Q = CV$$

For the circuit in Figure 1.18 the total charge in the circuit is:

$$Q = CV$$

where C is the total circuit capacitance and V is the total circuit voltage, therefore

$$C_{\text{Total}} = \frac{1}{\frac{1}{C1} + \frac{1}{C2} + \frac{1}{C3}}$$

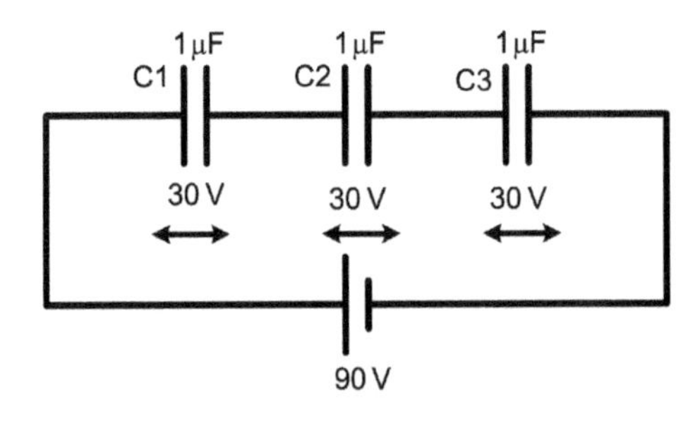

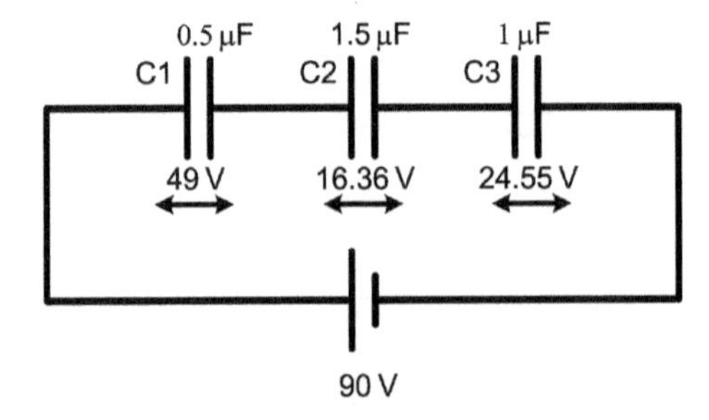

▲ **Figure 1.18** *Voltage sharing by capacitors*

so

$$C_{\text{Total}} = \frac{1}{\frac{1}{0.5} + \frac{1}{1.5} + \frac{1}{1}} \mu\text{F}$$

$$C_{\text{Total}} = \underline{0.27\,\mu\text{F}}$$

so

$$Q_{\text{Total}} = CV = 0.27\,\mu\text{F} \times 90\text{ V} = \underline{24.55 \times 10^{-6}\text{ C}}$$

This value is the charge (number of electrons) in the complete circuit, since it is a series circuit; it is the same charge in each of C1, C2 and C3!

$$\text{Since } Q = CV \text{ then } V_{C1} = \frac{Q}{C_1}$$

so

$$V_{C1} = \frac{24.55}{0.5} = \underline{49\text{V}}(\text{the powers cancel out})$$

similarly

$$V_{C2} = \frac{24.55}{1.5} = \underline{16.36\text{V}}$$

and

$$V_{C3} = \frac{24.55}{1} = \underline{24.55\text{V}}$$

To check that our answers are correct we add all three capacitor voltages together:

$$49\text{ V} + 16.36\text{ V} + 24.55\text{ V} = 89.91\text{ V}$$

which is approximately equal to 90 V, the applied voltage when taking into account rounding errors.

Example

In Figure 1.19, calculate:

The total circuit capacitance (1.084 uF).

The voltage across each capacitor (V_{C1}=1589.9 V, V_{C2}=3575 V, V_{C3}=1430 V).

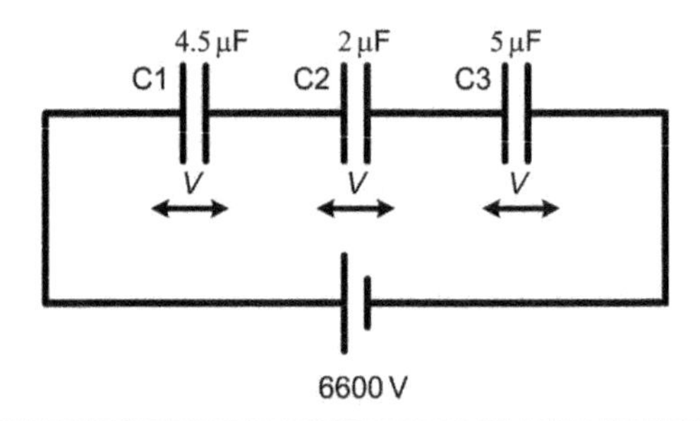

▲ **Figure 1.19** *Example – voltage sharing by capacitors*

Uses of capacitors

We can view the usage of capacitors in three broad categories:

1. DC Smoothing
2. AC Coupling
3. DC Blocking

Timing and filtering applications will be discussed later.

DC smoothing

Connecting a large value capacitor across a power supply output containing a non-smooth DC (direct current) output causes the DC output to become smoother (Figure 1.20).

The capacitor charges up during the positive half cycle (we pump electrons into the capacitor). After the peak of the positive half cycle has passed, and the voltage passing through the diode is reducing, the voltage within the capacitor is now higher than the

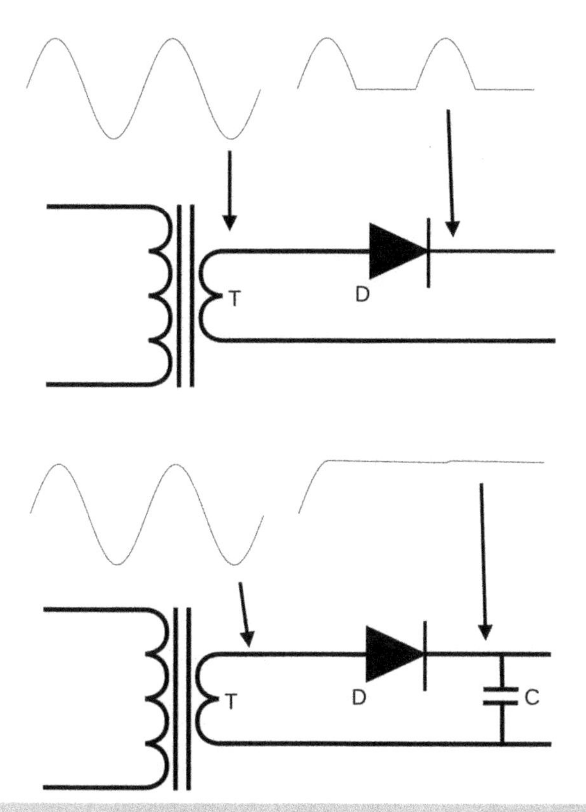

▲ **Figure 1.20** *Capacitance smoothing of PSU output*

diode voltage. The capacitor starts to provide current (electrons) to the load, making up for the falling voltage from the diode. A larger value capacitor will allow the output voltage to be smoother.

AC coupling, DC blocking

If when a voltage is applied across a capacitor the capacitor charges up (electrons flow into one plate, and flow out of the other plate), then during the charging time, electrical current is actually flowing through the capacitor. When a load is placed across the capacitor, the capacitor discharges (electrons return from the negatively charged plate to the positively charged plate) via the load. While the capacitor is discharging, electrical current is flowing through the capacitor. The above two paragraphs show that while the capacitor is charging and discharging, electrical current flows through it.

If we were to apply an AC (alternating current) voltage across the capacitor, we create the condition that allows alternating current to flow through the capacitor. This allows

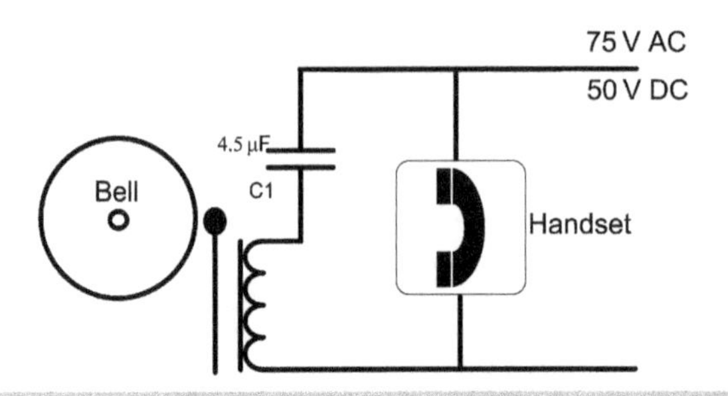

▲ **Figure 1.21** *AC blocking, DC coupling*

us the opportunity to use and pass through a circuit, AC voltages in the presence of DC voltages. In Figure 1.21, a telephone handset is connected across a 50 V DC supply from a telephone exchange. When an incoming call is instigated, a 75 V AC voltage is sent from the exchange to the telephone, to cause the telephone bell to ring. This AC signal is of sufficient size to cause the bell to ring. If the telephone bell coil were connected across the 50 V DC wires from the exchange, too much current would flow through the bell coil. To prevent this excess current flowing, a capacitor (C1) is connected in series with the bell coil, preventing DC current flowing through the coil, but allowing the AC ringing current to flow through it. We call this action DC blocking (it has blocked the 50 V DC from reaching the bell coil) or AC coupling (we have coupled the AC current to the bell coil).

Capacitance AC effects

The amount of current that passes through a capacitor does not relate to Ohm's law. The value of the current depends upon the AC voltage and the effective resistance of the capacitor. We know that the amount of charge (number of electrons) that can be stored in a capacitor is in proportion to the value of the capacitor. We know that DC voltages do not pass through a capacitor, and that AC current flows through the capacitor only while the capacitor is charging and discharging. The formula used for calculating the effective resistance that a capacitor presents to an AC voltage is:

$$\text{Reactance } (X_C) = \frac{1}{2\pi f C} \text{ Ohms } (\Omega)$$

where C is the capacitance in Farads and f is the frequency in Hertz. This effective resistance that a capacitor presents to an AC voltage is called reactance and measured in ohms. This formula tells us that this reactance X_C, is inversely proportional to both

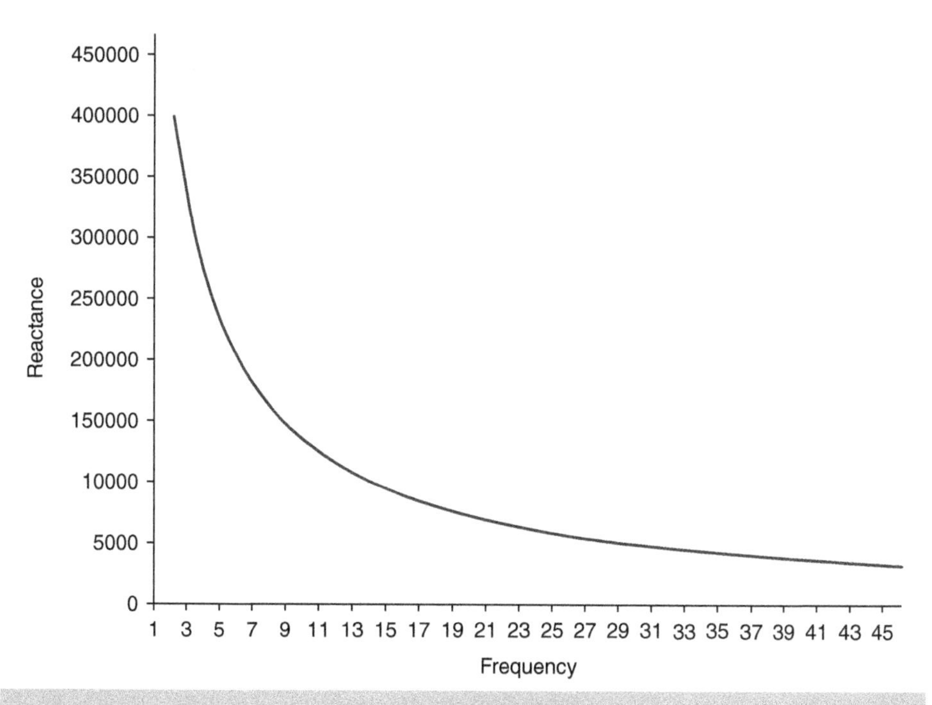

▲ **Figure 1.22** *Capacitive reactance with increasing frequency*

frequency and capacitance. The higher the frequency or capacitance, the lower the value of X_C. This effect can be seen in Figure 1.22.

Inductance

Inductance is the characteristic of an electrical circuit that opposes the starting, stopping, or a change in value of current. When a current passes through a wire, a magnetic field is created around the wire. When a moving/changing magnetic field cuts a wire, an electro-motive-force (EMF) or voltage is induced into the wire. The direction of EMF or voltage induced is such that it always opposes the force that created it. If a length of wire is wound round a former, the magnetic fields from each part of the wire add to each other, creating a greater magnetic effect. When the current increases or decreases through a coil of wire, the induced EMF or voltage is much greater than if the wire were not wound into a coil. Inductance is the property that the coil possesses, which opposes the increase or decrease in current flow due to the induced EMF or voltage opposing the change in current. The amount of EMF or voltage induced into a coil of wire depends upon:

- number of turns of wire;
- rate of change of magnetic field.

Lenz's law says the induced EMF always opposes the change producing it, leading to terminology such as 'back EMF'.

Faraday's first law states that whenever a conductor cuts a magnetic field or vice versa an EMF is induced in it and it is in such a direction as to oppose the cause of it.

Faraday's second law states that the magnitude of induced EMF is equal to the rate of change of flux linkage.

Inductors, due to the laws mentioned above, oppose the change of current flowing through them. They oppose an increase in current flow, and they oppose a decrease in current flow. They do not oppose a constant current. The amount of opposition a coil gives to a changing current depends upon the number of turns that the coil possesses and the rate of change of the current. The greater the number of turns, the greater the opposition, the faster the current tries to change, the greater the opposition. This is summed up in the following formula:

$$e = -N\frac{\mathrm{d}\Phi}{\mathrm{d}t}\ \text{Volts (v)}$$

where e is the induced EMF, N is the number of turns of wire and $\frac{\mathrm{d}\Phi}{\mathrm{d}t}$ is the rate of change of flux.

The minus sign shows us that the induced EMF or current is set up in a direction so as to oppose the cause of it. Inductance is typified by the behaviour of a coil of wire in resisting any change of electric current through the coil. Arising from Faraday's law, inductance L may be defined in terms of the EMF generated to oppose a given change in current. If the rate of change of current in a circuit is 1 Amp/s and the resulting electromotive force is 1 V, then the inductance of the circuit is 1 H.

$$\text{Inductance} = \frac{\text{Volts/Seconds}}{\text{Current}}\ \text{Henries}\ (L)$$

When AC voltages are applied to an inductor, the amount of opposition to current change (Lenz's law) depends upon the value of the inductance and the frequency of the

AC voltage. This opposition to current flow in inductors fed by an AC voltage is similar to resistance but called reactance. The formula for the reactance of an inductor is:

$$\text{Reactance } (X_L) = 2\pi f L \text{ Ohms } (\Omega)$$

where L is the inductance in Henries and f is the frequency in Hertz.

This formula shows us that a higher frequency AC signal would see an inductor as a higher value resistance compared to a lower frequency signal. Similarly, an AC signal would experience a higher resistance with a large value inductor compared to a lower value inductor.

This effect can be seen in Figure 1.23.

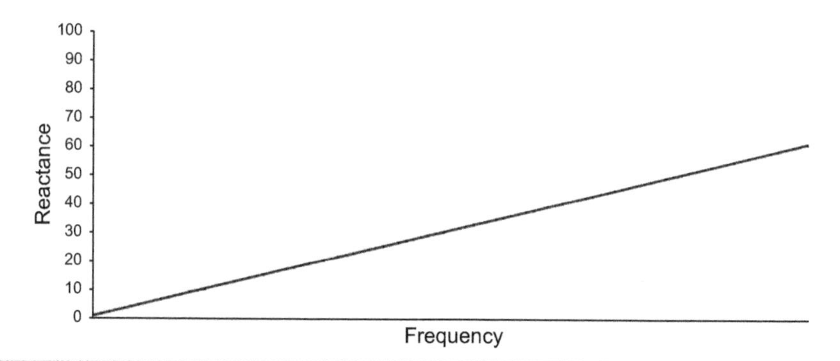

▲ **Figure 1.23** *Inductive reactance with increasing frequency*

Uses of inductance

Due to the effect of reactance increasing with increasing frequency, inductors can be used as filters. They allow DC to pass through unimpeded, but to higher frequency AC signals they appear as high value resistances, therefore, inductors filter out high frequencies or harmonics.

Inductive coupling allows the magnetic field from one inductor to be coupled or transmitted to an adjacent inductor. This allows a device called a transformer to work. See section on transformers.

It is the reactance of transformer windings that limits the flow of AC current into a transformer when it has no load. A DC voltage applied to a transformer would cause a very large DC current to flow through the winding, limited only by the DC resistance of the winding. For this reason, we cannot apply DC to a transformer.

Semiconductors

Diode effect

Semiconductor materials have revolutionised electronics. The most popular material to use in a semiconductor is silicon (Si). A single crystal of silicon, ingot in shape, is purified by locally heating parts of the ingot, moving the heat zone slowly down the ingot. As the heated silicon cools, impurities in the silicon remain in the molten zone. The molten zone slowly moves towards the base of the ingot, taking the impurities with it. This way, the silicon is purified to a very high level, 99.99% pure. Once the silicon has been purified it is a poor conductor of electricity, a semiconductor. This is due to the number of electrons in the outer shell of the silicon. To make the pure silicon useful, impurities are added to the pure silicon to give us controlled amounts of conductivity. If pure silicon has a small amount of doping material added, say antimony, arsenic or phosphorous, then the additional electrons in the outer shell (valence electrons) cause the silicon to have excess electrons and become a good conductor. If pure silicon has a small amount of doping material added, say boron, aluminium or gallium, then the additional atoms of the dopant cause a shortage of electrons in the outer shell (valence electrons) causing the silicon to have a shortage of electrons and become a good conductor.

A shortage of electrons or an excess of electrons gives the opportunity of electron movement (current flow) when a voltage is applied. Missing electrons are called 'holes', as they represent a place where an electron could fit in. Silicon with excess negative electrons is called 'N' type material, and silicon with an excess of positive holes is called 'P' type material (P for positive, N for negative). When a section of 'P' type silicon is in contact with 'N' type silicon, in a small region, where the two materials touch, excess negative electrons and the excess holes merge and combine. The negative charge of the electron and the positive charge of the hole (missing electron) cancel out (Figure 1.24).

This small region where the holes and electrons have combined is now unable to carry any electrical current; there are no free electrons or holes. It is depleted of current carriers. This area or zone is named the depletion layer. Two types of silicon joined together (P and N type) are termed a PN junction. A PN junction possesses unusual characteristics. Consider a battery is connected across a PN junction, with the positive

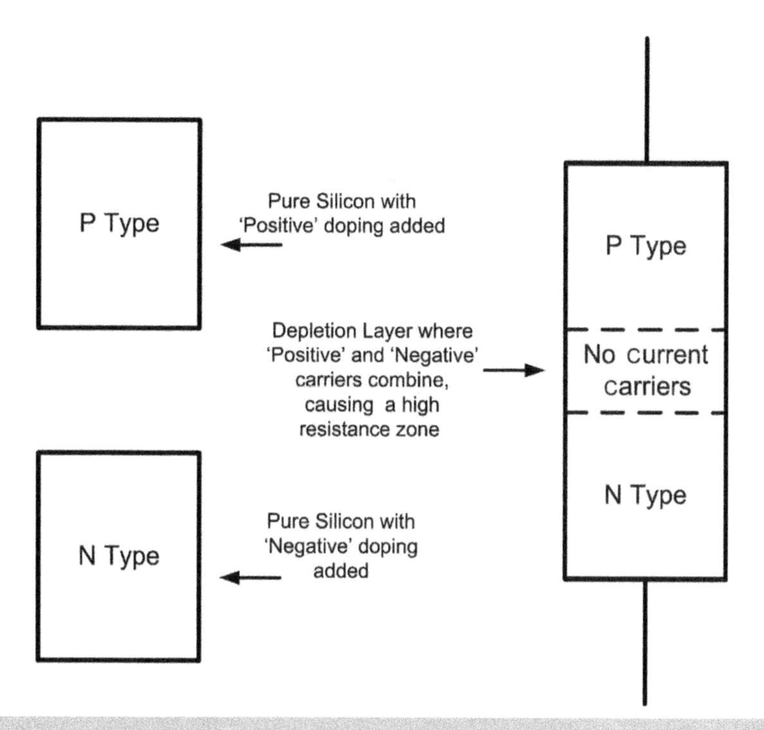

▲ **Figure 1.24** *PN junctions*

terminal of the battery connected to the P end of the PN junction, the negative terminal of the battery connected to the N end of the PN junction. With the battery set to produce 0.1 V, the depletion layer, consisting of an area of zero current carriers, acts as an insulator, preventing much, if any, current from flowing.

As the battery voltage is increased, a very small current may flow. As the battery voltage is increased, at about 0.6–0.7 V, the electron pressure has reduced the width of the depletion layer sufficiently to allow electrons to 'jump' across the layer and return to the battery. The response of a PN junction with the positive terminal of a battery connected to the P end of the junction is very typical and is shown in Figure 1.25.

With the voltage polarity connected this way, the PN diode is said to be forward biased. If the battery terminals were now reversed, starting with a low battery voltage, we would find that only a very small current may flow through the PN junction. As the battery voltage is increased, more free electrons are being dragged out of the N part of the PN junction; more electrons are being injected into the P end of the PN junction. These events increase the size of the depletion layer, reducing the likelihood of any current flow. If the battery voltage is increased above a certain level, the electrical stress across the depletion layer becomes too great, and the PN junction breaks down. Once the PN

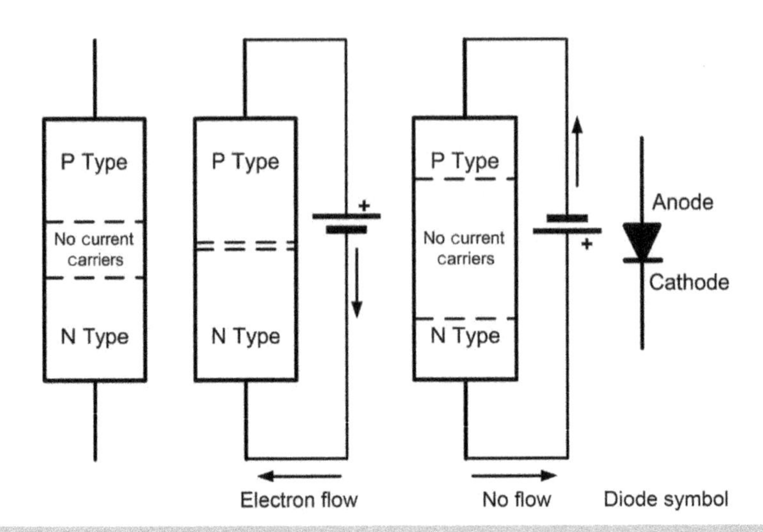

▲ **Figure 1.25** *PN junction forward and reversed biased*

junction depletion layer has broken down, it becomes a near short circuit, allowing the battery to force large currents through the junction. Without a series resistor to limit the current, the PN junction would melt within milliseconds. When a PN junction is connected in this direction, it is said to be reverse biased. These two characteristics give the PN junction it's most useful function: it allows current to flow in one direction but prevents current flow in the other. It is called diode action. We use diode action to convert AC to DC (rectification) and detect radio signals. Diodes, when they are operated in the forward biased direction, will exhibit a voltage across them of approximately 0.6–0.7 V. The amount of voltage measured depends upon the internal construction of the diode and how much current is passing through the diode. It is the diode that determines the voltage. Diodes are rated to withstand certain levels of forward biased current, for example, 1 A, 100 A, etc. They also have a power rating in watts, for example, 400 mW, 10 W, etc. Exceeding stated limits will damage the diode. Diodes are rated to withstand certain voltage levels when reverse biased. A diode suitable to rectify 230 V AC would typically be rated at 400 V peak reverse voltage. Meaning that, as long as the mains voltage did not exceed 400 V, the diode will not reach its breakdown voltage of 400 V. When specifying diodes, these three parameters need to be taken into account. Other parameters exist which can be found in manufacturers' data sheets.

If a diode is reverse biased with a voltage above its reverse breakdown voltage, it will break down and suffer a catastrophic current flow. If we connect a resistor in series with the diode, and the value of this resistor is chosen to allow a safe current to flow through the diode, then the diode will sit happily with the breakdown voltage present at its terminals. This controlled breakdown effect is called the Zener effect. Any diode can be used as a Zener diode, but diodes used for power rectification are always operated below the Zener breakdown voltage, and Zener diodes are always operated at a voltage that is higher than their Zener voltage via a series resistor (Figure 1.26).

The choice of resistor value depends upon the value of current that you wish to flow through the Zener diode (Figure 1.27). If you have a 10 V supply, a 6.3 V Zener, and you want the Zener to carry 10 mA, then the resistor will have 10 V – 6.3 V = 3.7 V dropped across it. With 10 mA required to flow, $R = \frac{V}{I}\ \Omega = \frac{3.7}{0.01} = \underline{370\Omega}$.

Zener diodes have power ratings. The higher the current passing through the Zener or the higher the voltage across the Zener, the higher the power dissipation. The power rating (P = VI W) should not be exceeded or the Zener will be destroyed.

Zener diodes are available in a series of preferred voltage values: 5.6 V, 9.6 V, 15 V, etc. and a range of power dissipation ratings: 400 mW, 10 W, etc.

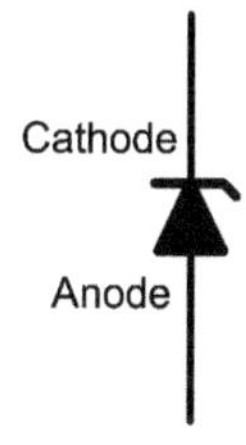

▲ **Figure 1.26** *Zener diode symbol*

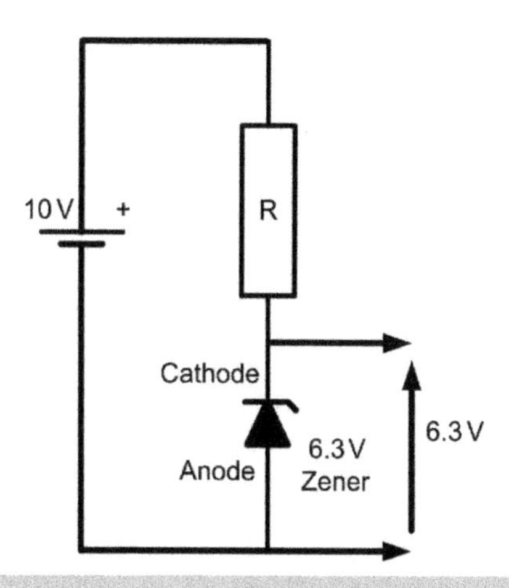

▲ **Figure 1.27** *Typical zener diode circuit*

Light emitting diodes

If a diode (PN junction) is created using a different set of doping elements such as Gallium and Arsenic, then when the diode junction is forward biased, the junction emits light. The wavelength of light and the amount of light emitted depends upon the choice of doping elements and the amount of doping elements used. Since the light emitting diode (LED) is forward biased when in use, it requires a series resistor to limit current flow. Typical values of the junction voltage from an LED when emitting light is between 2 and 3 V. We decide the current flow we require through the diode to give us the required amount of emitted light, then we calculate the value of the series resistor by looking up the manufacturer's data sheet for the value of the forwards voltage. The diode controls the forward volts drop, we specify the current and control the current with the series resistor. Typical small LEDs operate at 20 mA. LED technology is advancing rapidly; the goal of replacing incandescent lamps with cheap and efficient LEDs is within reach (Figure 1.28).

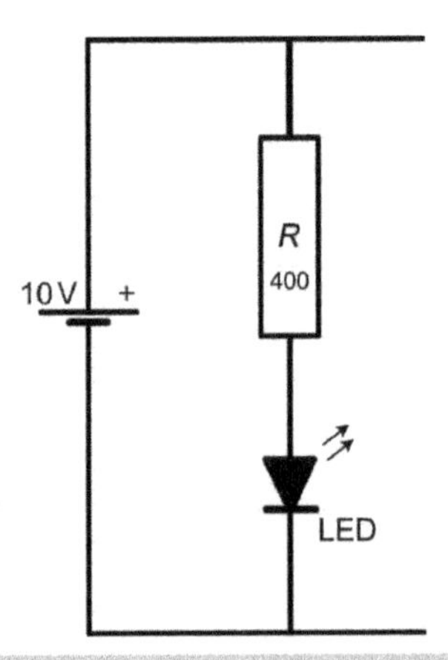

▲ **Figure 1.28** *Typical LED circuit*

Opto sensitive diodes

PN junctions, with suitable quantities of the appropriate doping material will be light sensitive. With the diode reverse biased and a minute amount of current flowing, when the junction is exposed to light, the amount of current flowing increases in proportion to the amount of light received (Figure 1.29). These devices are called opto diodes or photo diodes. By changing the quantity and type of doping elements and structure, solar cells can be made which generate electricity when exposed to light.

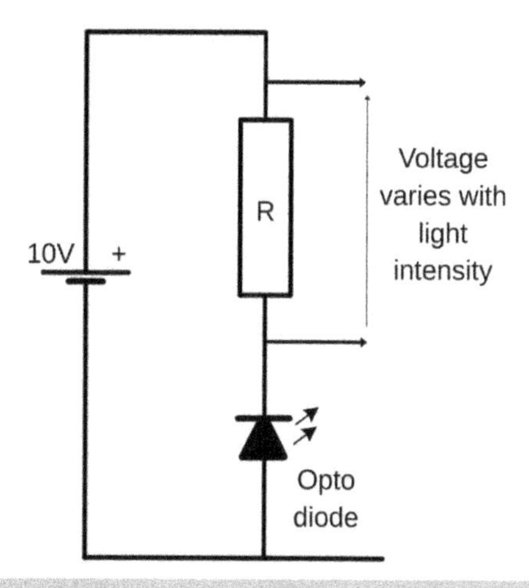

▲ **Figure 1.29** *Typical opto diode circuit*

Transistors

A transistor is a semiconductor device that has two PN junctions (Figure 1.30). If the slab of silicon is manufactured to contain two PN junctions close to each other (microns apart) additional characteristics of multiple PN junctions become available. The polarity of the silicon is alternated through each layer, that is, N type, P type and N type; or P type, N type and P type.

Usually referred to as NPN or PNP. Each layer has an electrode attached; the two outer connections are called emitter and collector, and the central connection is called base.

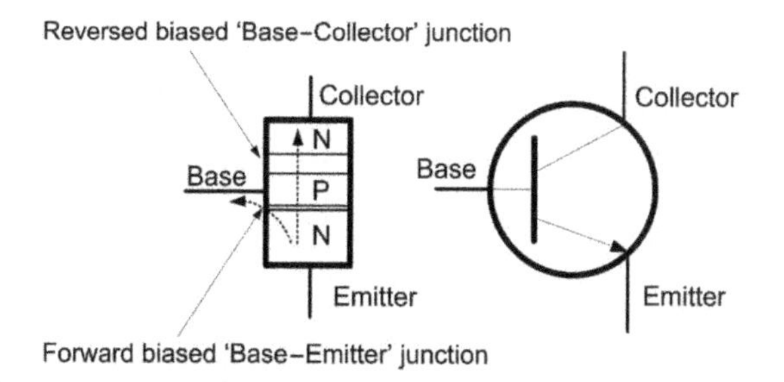

▲ **Figure 1.30** *Transistor junctions and symbol*

In a circuit consisting of a transistor (a two PN junction device) the two resistors are connected as shown in Figure 1.31. The PN junctions and depletion layers can be seen in Figure 1.32. With a high positive voltage at the top of the circuit, electrons want to travel from the bottom of the circuit, through the transistor, via its emitter, through the base, to the collector. Unfortunately, the transistor has two depletion layers, the emitter/base PN junction, and the base/collector PN junction. If a small current is caused to flow through the base terminal, the emitter/base junction becomes forward biased and electrons can pass through this layer easily. The base/collector junction is reverse biased, so impeding the passage of electrons to the collector. The

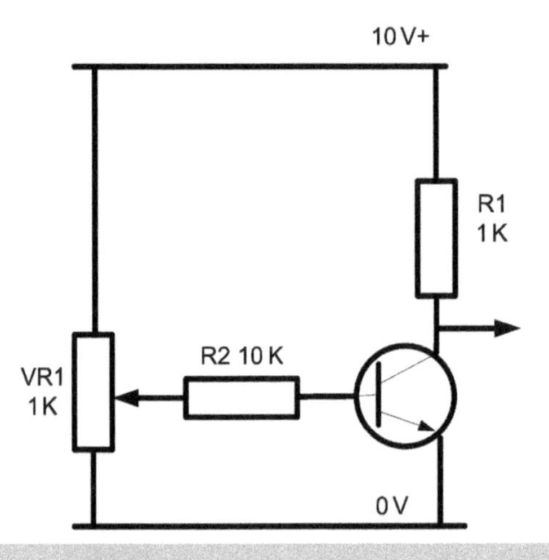

▲ **Figure 1.31** *Simple transistor circuit*

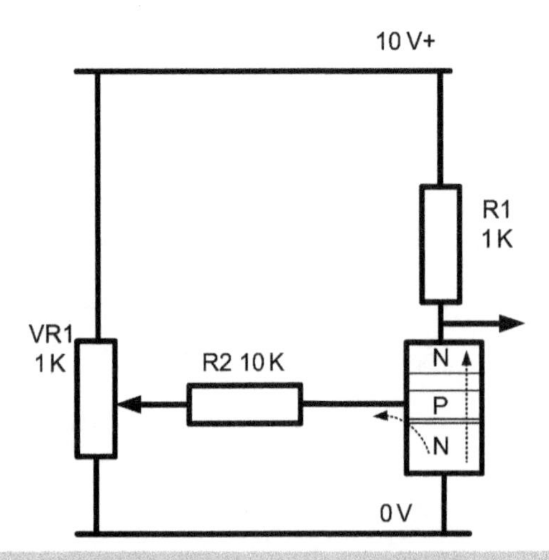

▲ **Figure 1.32** *Simple transistor circuit with depletion layers*

combination of the two depletion layers being very close to each other (microns apart) with the emitter/base junction being forward biased and the high voltage stress across the device causes electrons to flow from the emitter to collector. The amount of current that flows through the collector is directly related to how much current flows through the base terminal.

Field effect transistors

In a field effect transistor (FET), the current carrier is of one type only, electrons for N type or holes for P type silicon. The FET is constructed from one piece of silicon, with current flowing from end to end, without any depletion layers to negotiate. All FETs work by restricting the width of the channel of silicon that the current flows through. Just like a wire conductor, reduce the CSA of a conductor and its resistance increases. There are two basic types of FET: junction FET (JFET) and insulated gate FET (IGFET). A variety of sub-groups define different FETs (Figure 1.33).

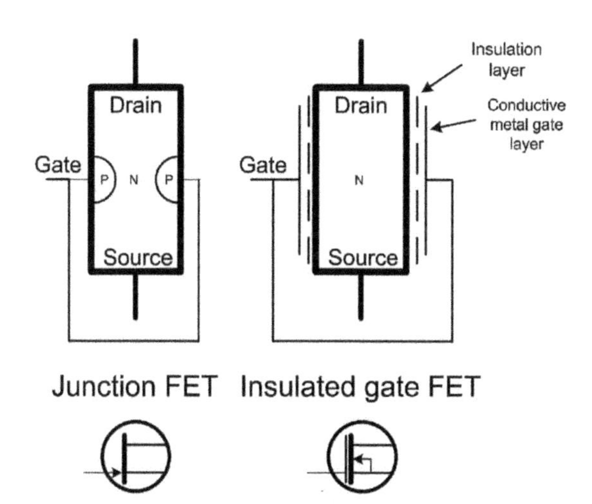

▲ **Figure 1.33** *Junction FET and insulated gate FET with symbols*

Junction FET

The JFET works by restricting the width of the current carrying the silicon channel by positioning PN junctions on either side of the channel. By increasing the width of the depletion layer, the remaining width of the channel is reduced, increasing the resistance of the channel and reducing current flow.

Insulated gate FET

The IGFET does not have any PN junctions. The silicon channel has two conductive plates on either side, insulated from the silicon channel via a microscopically thin layer of insulation. The insulation is usually made from a metal oxide of silicon. When a voltage is placed onto the two conductive plates, an electric field is produced between the two plates, restricting the width of the silicon channel, increasing its resistance and reducing current flow. IGFETs have very high input impedance due to the insulated layers. This high input impedance can make IGFET devices very prone to static damage. Depending upon the doping elements and quantities, an IGFET can be an enhancement mode or depletion mode device. An enhancement IGFET with zero volts on its gate will not conduct or be off, a depletion IGFET with zero volts on its gate will conduct or be on. This leads to a unique configuration for IGFETS.

Complementary metal oxide silicon (CMOS)

In Figure 1.34 two IGFETs are shown with their gates connected together. One IGFET is an N-channel device, the other P-channel. As both gates are tied together, they receive the same voltage. As one device is an enhancement type and the other a depletion type, only one IGFET will conduct at any one time. This means that there is never a

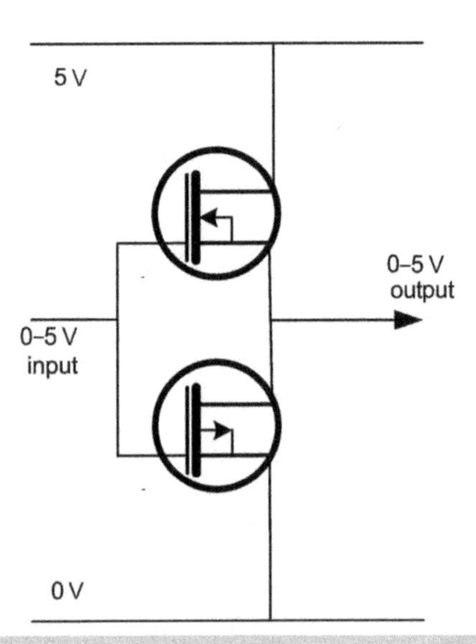

▲ **Figure 1.34** *CMOS output stage*

complete path between the ground line and +5 VCC. Therefore any current travelling between these two devices will be extremely small. A small voltage on the gates can control both IGFETs, but the current consumption is very small. When current flows in the gate circuit, it is used to charge up the capacitance that exists between the gate and the channel, separated by the very thin layer of insulation which acts like a dielectric. As you increase the frequency of the input voltage switching from 0 to 5 V, the gate capacitance is charged and discharged more frequently, causing an increase in current consumption. This is why higher clock speeds increase the current, power and heat dissipated by a modern computer. Millions of these transistor pairs can be connected together to create logic and computing elements.

Signal Shaping

DC restoration

DC restoration involves the translation of the embedded DC level of an AC signal to a different DC level. If you had a signal that consisted of a sine wave, levels +5 and −5 V, its average voltage level is 0 V, this signal could be translated (or clamped) to 0 V. In Figure 1.35 we can see that the lower part of the waveform which was at −5 V is now clamped to approximately 0 V, the remainder of the waveform extending up to +10 V.

In Figure 1.36, we can see that the upper part of the waveform which was at +5 V is now clamped to approximately 0 V, the remainder of the waveform extending down to −10 V.

We can change the level of the output signal even further if we were to introduce an additional source of voltage to increase the voltage shift. In Figure 1.37, we can see a

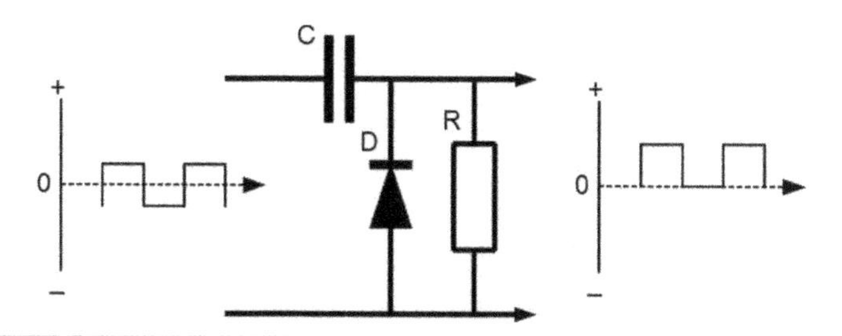

▲ **Figure 1.35** *Clamped to zero, positive going*

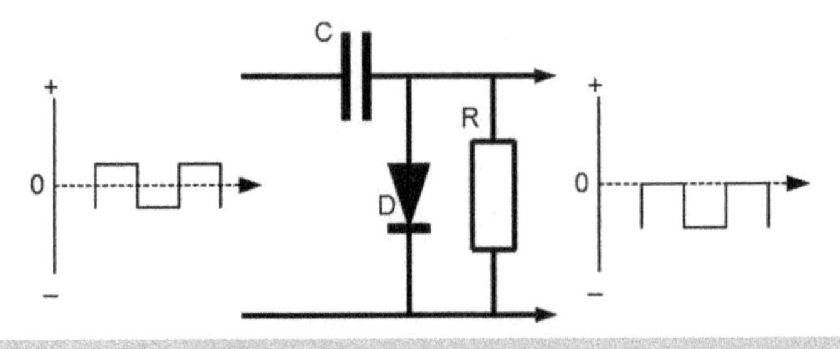

▲ **Figure 1.36** *Clamped to zero, negative going*

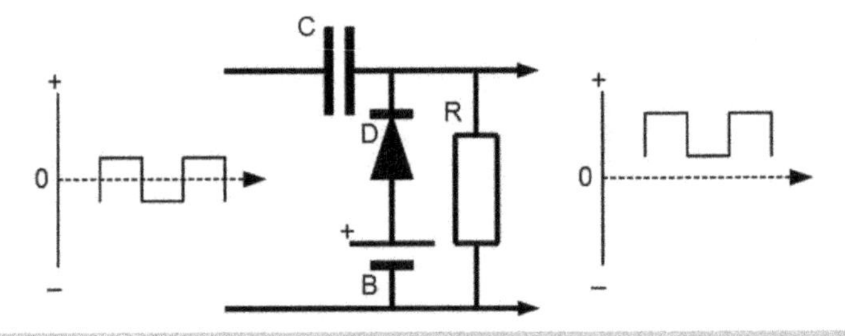

▲ **Figure 1.37** *Clamped to positive value, positive going*

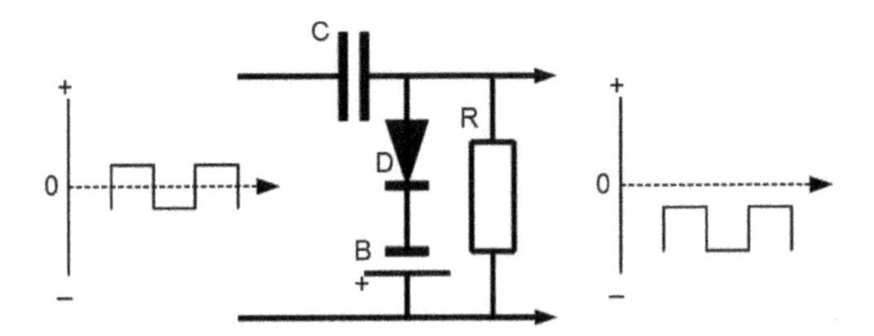

▲ **Figure 1.38** *Clamped to negative value, negative going*

battery is introduced to increase the amount of voltage offset. The additional amount offset is equal to the battery voltage.

We can offset the voltage in either a positive or negative direction (Figure 1.38).

DC restoration is used occasionally to restore a signal level after transmission via a long line. Typical applications can be found in Radar, TV and video systems where specific voltage levels are required for specific parts of an analogue video signal. Having transmitted a video signal over a cable, through capacitors and transformers, it is necessary to restore the DC level to make the video signal usable. The DC clamp effect is also used by voltage multipliers.

AC limiting and squaring

Sometimes AC signals, when being received from a dynamic source (microphone, amplifier, etc.), can vary in amplitude to such an extent that overloading can occur in connected equipment. To prevent this, AC limiting can be used. AC signal limiting can be implemented in a variety of ways; we will consider diode limiting (Figure 1.39).

To obtain an output signal with a shape approaching that of a square wave, the input signal can be increased, so the output signal utilises only a small part of the input signal.

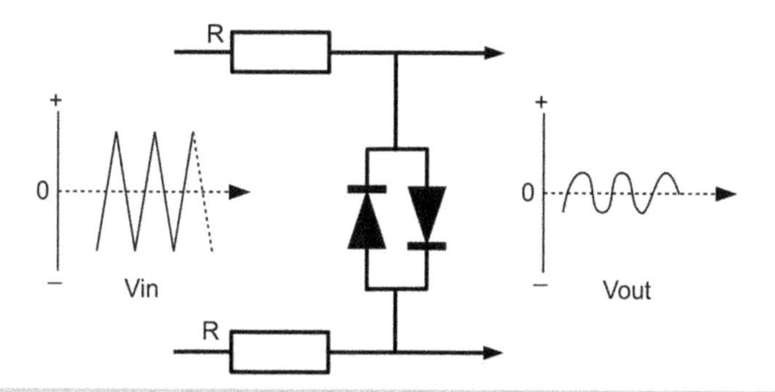

▲ **Figure 1.39** *Back to back diodes used to clip AC signal*

Diode limiting

Small signal diodes can be connected 'back to back' to limit the amplitude of an AC waveform to approximately ±0.65 V.

Schmitt trigger circuits

Some electronic devices, specifically digital gates, require clearly defined input signals, either logic high or logic low. Unfortunately, real world signals, especially those from the analogue world or from human input, typically are slowly changing from one level to another. If a slowly changing signal is fed into a digital device, the digital device will assume the slowly changing input signal, for a short period, is a rapidly changing digital signal, causing the output of the digital device to miss operate. To help overcome this problem, Schmitt input devices were invented. Schmitt triggers add hysteresis to a circuit. This means that the input signal has to exceed the normal switching voltage to be registered as a valid changing input signal. When the input signal reverts to the original level, the opposite happens; the input signal has to exceed the normal switching threshold before the signal is registered. If the degree of hysteresis is sufficient, slow changing analogue signals can be input to a digital gate and the gate will switch with safety, without the output chattering as the input traverses the normal switching threshold. In Figure 1.40, a digital gate is shown with its input connected to a resistor capacitor CR time constant circuit. When the circuit is powered up, the capacitor slowly charges up via the resistor. The slowly rising voltage is applied to the gate input at A, see Figure 1.40. The digital gates output will switch to logic one or zero depending upon the voltage applied to the input. The CR time constant provides a slowly changing input voltage. When the level of this voltage reaches the switching point (from zero to one) the digital gate can incorrectly evaluate the voltage applied to the gate. During this period of uncertainty, the output of the gate will rapidly switch between logic zero and one. This can be seen in Figure 1.40. When the digital device is replaced by one containing a Schmitt trigger, the response looks like Figure 1.40, where the output X switches just once when the input at X slowly rises. Figure 1.40 shows the relationship between the input and output of a Schmitt trigger circuit.

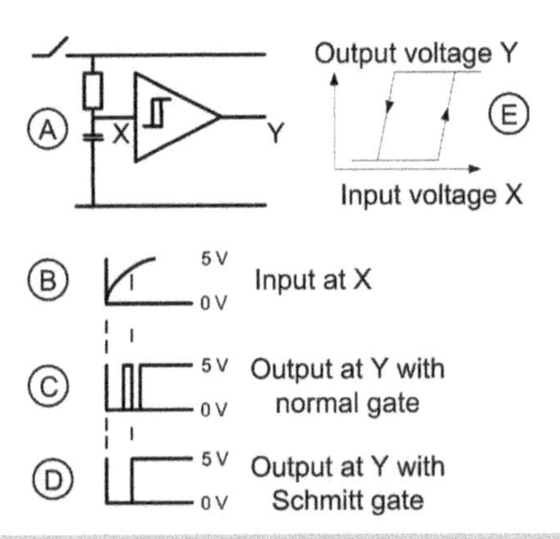

▲ **Figure 1.40** *Schmitt trigger stops slow input causing problems*

Operational Amplifiers

Simple op-amp

Operational amplifiers are small electronic devices that possess unlimited gain (Figure 1.41).

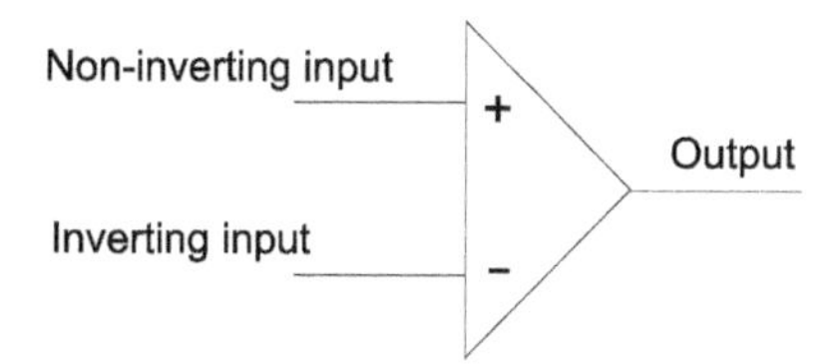

▲ **Figure 1.41** *Op-amp symbol*

Operational amplifiers (op-amps) typically have two inputs and one output. One input, named the non-inverting input or '+' possesses very high gain and the output pin follows in phase with this input. The other input, named inverting input or '– ' has the identical gain to the '+' input but the phase of the signal is inverted between this input and the output. Op-amp theory relies upon op-amps possessing infinite gain and infinite input impedance. Assuming an op-amp possess these two characteristics then any very small voltage applied to either of the two inputs would cause, due to infinite gain, the output pin voltage to move immediately to the voltage limits of its travel. Typically, op amps are connected to two power supply voltages, for example, +5 and –5 V or +15 and –15 V. In this case the output of the op amp would reach one of these voltage rails. If we connect the op amp as shown in Figure 1.42 any change in voltage at the output pin is fed back to the op amp input pin via resistor $R_{feedback}$ (R_f). As the input signal has to pass through the input resistor, R_{input} (R_i), the ratio of these two resistors (R_f and R_i) determines the gain of the complete circuit, for example if both R_f and R_i were both 1,000 Ω, the voltage dropped across R_f and the voltage dropped across R_i would be equal, only when the op amp output voltage was equal to the input voltage and the voltage at the junction of R_f and R_i was approximately zero. If the voltage at the junction of R_f and R_i was not approximately zero, the output voltage would shoot to one of the voltage rails! In Figure 1.42, the input voltage is connected to the inverted input of the op amp, so every voltage output by the op amp will be inverted.

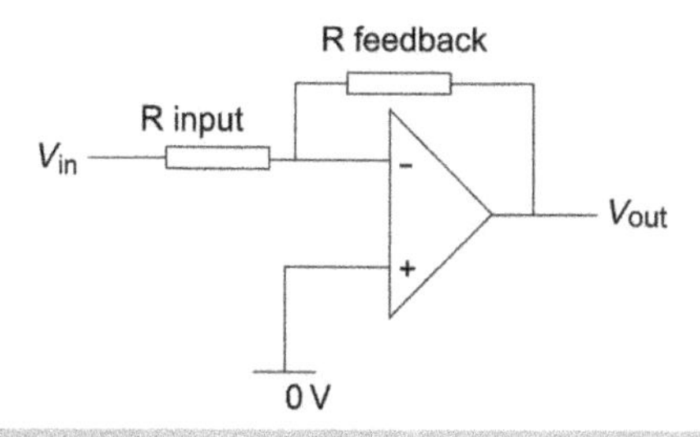

▲ **Figure 1.42** *Op-amp with input and feedback resistors*

For example,

+1 V input = –1 V output

–2 V input = +2 V output

As the op-amp is a powered device (connected to power supply rails) it has the power to force the op-amp output pin to assume any voltage within its power supply range. The op-amp always reaches equilibrium when the input voltage and the output voltage both pass through R_i and R_f respectively and maintain approximately zero volts at the input to the op-amp. For this reason (the op-amp signal input is always near zero volts), the op-amp input is called a virtual earth. We can alter R_f or R_i to change the gain of the amplifier.

If R_f or R_i were changed in value, the gain of the complete amplifier would change.

To calculate the gain of an op-amp based amplifier we divide the feedback resistance value by the input resistance value.

For example, if both the resistors = 1,000 Ω in value,

$$\text{Gain} = \frac{-R_f}{R_i}, \quad \text{Gain} = \frac{-1{,}000}{1{,}000} = \underline{-1}$$

If $R_f = 2\ \text{k}\Omega$ and $R_i = 500\ \Omega$ then:

$$\text{Gain} = \frac{-R_f}{R_i}, \quad \text{Gain} = \frac{-2{,}000}{500} = \underline{-4}$$

If $R_f = 500\ \Omega$ and $R_i = 2\ \text{k}\Omega$ then:

$$\text{Gain} = \frac{-R_f}{R_i}, \quad \text{Gain} = \frac{-500}{2{,}000} = \underline{-0.25}$$

In all of the above cases, as the input signal is connected to the inverting input '–', then a '–' should precede any gain figure, for example, –4, –2 or –0.25.

Multiple input op-amp amplifier

If we were to construct a circuit such as Figure 1.43, a summing amplifier would be created. A summing amplifier has two or more inputs. Each input has its own input resistor and shares the common feedback resistor.

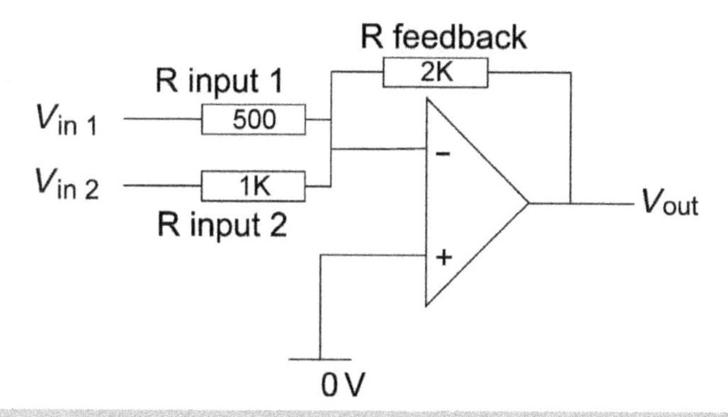

▲ **Figure 1.43** *Summing amplifier*

The gain calculation for a summing amplifier is as follows:

To calculate the voltage out, treat each signal path separately and algebraically add the separate output voltages together.

Through signal path 1 $(R_{in1}, R_f) = \text{Gain} = \dfrac{-R_f}{R_{in1}}$

Note the minus sign due to the use of the inverting input.

Through signal path 2 $(R_{in2}, R_f) = \text{Gain} = \dfrac{-R_f}{R_{in2}}$

The output voltage V_{out} is equal to (input$_1$ Gain$\times -V_{in1}$) + (input$_2$ Gain $\times -V_{in2}$) …

Always remember that when feeding signals into the inverting input, to invert each individual signal.

For example, in Figure 1.43 $R_{in1} = 500\Omega$, $R_{in2} = 1{,}000\ \Omega$ and $R_f = 2{,}000\ \Omega$.

Gain for signal path $R_{in1}, R_f = -\text{Gain} = \dfrac{R_f}{R_{in1}} = \dfrac{2{,}000}{500} = \underline{-4}$

Gain for signal path $R_{in2}, R_f = -\text{Gain} = \dfrac{R_f}{R_{in2}} = \dfrac{2{,}000}{1{,}000} = \underline{-2}$

To calculate the voltage at V_{out}, we use the voltage applied to an input multiplied by the gain of that signal path.

$V_{out} = V_{in} \times \text{Path gain}$

In Figure 1.44 a five input summing amplifier is shown. It is easier to use the table method to calculate the final output voltages if all input voltages and the resistor values are known.

Referring to Table 1.3 we calculate the gain of each input to output path (R_f divided by each R_{in} in turn), multiply each input voltage by each gain, taking note of any minus signs, sum all the voltages together, taking note of any minus signs.

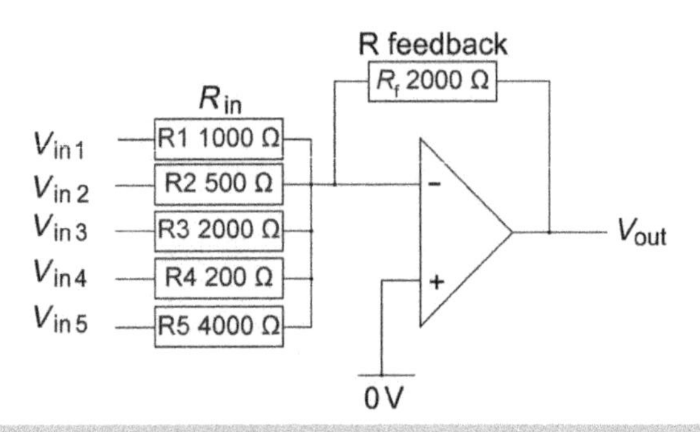

▲ **Figure 1.44** *Five input summing amplifier*

Table 1.3 *Five input summing amplifier*

	Input resistor	**Feedback resistor**	**Gain**	V_{in}	V_{in} **× Gain**
	(R_{in}) Ω	(R_f) Ω	$\frac{-R_f}{R_{in}}$	**V**	**V**
in1	$R_{in1} = 1K$	2K	–2	0.1	–0.2
in2	$R_{in2} = 500$	2K	–4	–0.5	2.0
in3	$R_{in3} = 2K$	2K	–1	–0.1	0.1
in4	$R_{in4} = 200$	2K	–10	0.33	–3.33
in5	$R_{in5} = 4K$	2K	–0.5	–0.2	0.1
Output					–1.33 V

Wien bridge oscillator

So far we have connected a signal to the inverting input of an op-amp (–). We have also connected our feedback resistor to the inverting input. This style of connection ensures negative feedback with stable operation.

An alternative configuration can be used to create a non-inverting amplifier. This is shown in Figure 1.45. In this configuration, the input signal is connected to the non-inverting input (+) and the feedback resistor is connected to the inverting input (–) ensuring controlled negative feedback.

The formula for the gain of an op-amp circuit when connected this way is:

$$\text{Gain} = \frac{V_{OUT}}{V_{IN}} = \frac{R_{f1}}{R_{f2}} + 1$$

If we were to flip the op-amp, so that the minus sign was at the bottom of the amplifier symbol, and repositioned the resistors accordingly, we would see a circuit similar to Figure 1.46.

The gain is still calculated with the formula:

$$\text{Gain} = \frac{V_{OUT}}{V_{IN}} = \frac{R_{f1}}{R_{f2}} + 1$$

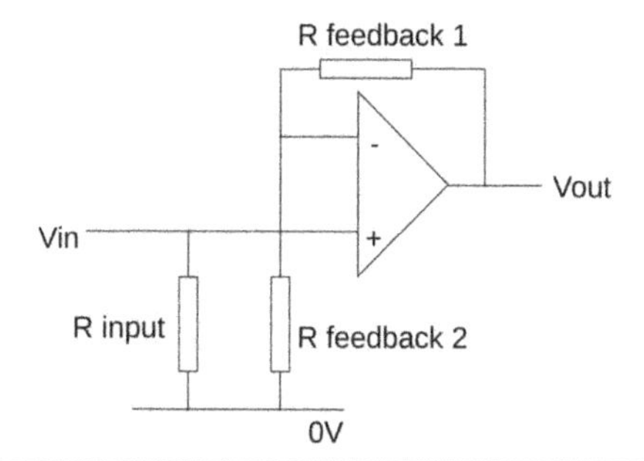

▲ **Figure 1.45** *Non-inverting amplifier*

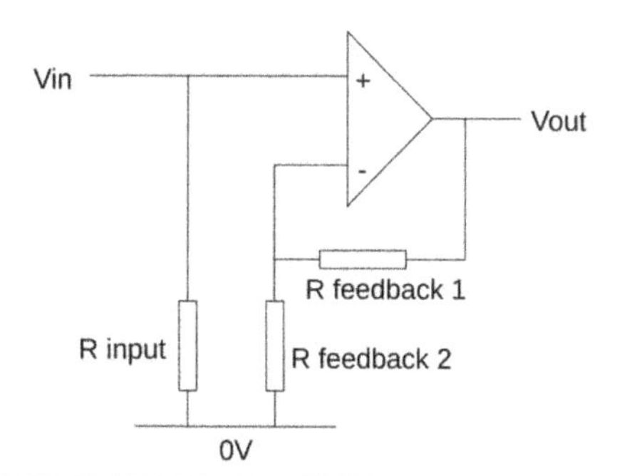

▲ **Figure 1.46** *Non inverting op-amp – standard layout*

Referring to Figure 1.49 we have added an extra connection from the output of the op-amp to the non-inverting input (+).

Disregarding C_1 and R_1, a connection from an output to a non-inverting input will cause positive feedback. (Positive feedback occurs when a small amount of an amplifier output voltage is sent to the amplifier input, the amplifier gain amplifies this small signal, and it is sent around again. Within a fraction of a second, the size of the signal will reach a maximum.)

We have installed in-line with our feedback connection, C_1 and R_1. Remember that capacitors are frequency sensitive. Assuming that a small AC signal over a range of frequencies is produced by the amplifier, these signals can pass through C_1 and R_1. C_2 and R_2 are connected from the amplifier input to 0 V. C_2 is also frequency sensitive. The formulae for capacitive reactance is:

$$\text{Reactance } (X_C) = \frac{1}{2\pi fC} \text{ Ohms } (\Omega)$$

The effective ohmic value of C_1 and C_2 varies with frequency. C_1, in series with the feedback signal, allows higher frequencies to pass through. C_2 partially connects the feedback signal to 0 V (earth), which tends to 'short out' higher frequencies to earth. There will be one frequency where the capacitor resistor combination appears like a bandpass filter as shown in Figure 1.47. The frequency can be calculated with the formula: $f = \frac{1}{2\pi RC}$ where R and C are the values of either resistor or capacitor as $R_1 = R_2$ and $C_1 = C_2$.

The result of sweeping a frequency into the input of this bandpass filter would give a voltage output similar to that shown in the lower part of Figure 1.48. The upper part of Figure 1.48 displays the phase shift between the output and input of the filter.

The voltage output from the resistor capacitor filter is 33.33% of the input at the target frequency. The phase of the output signal is 0° at the target frequency. The change in gain and phase, combined with the in-phase feedback from the op-amp, give the perfect conditions to allow the complete circuit to oscillate. Due to the voltage loss

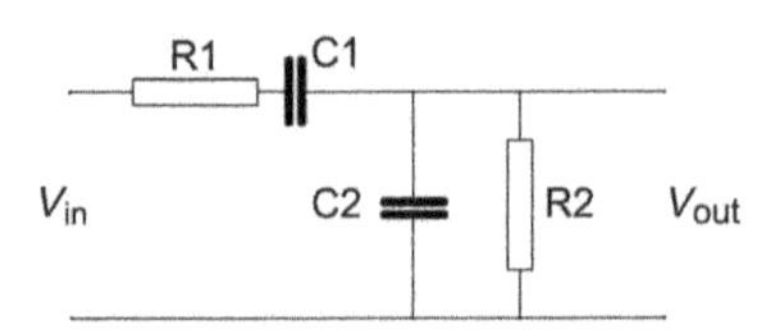

▲ **Figure 1.47** *Bandpass filter circuit*

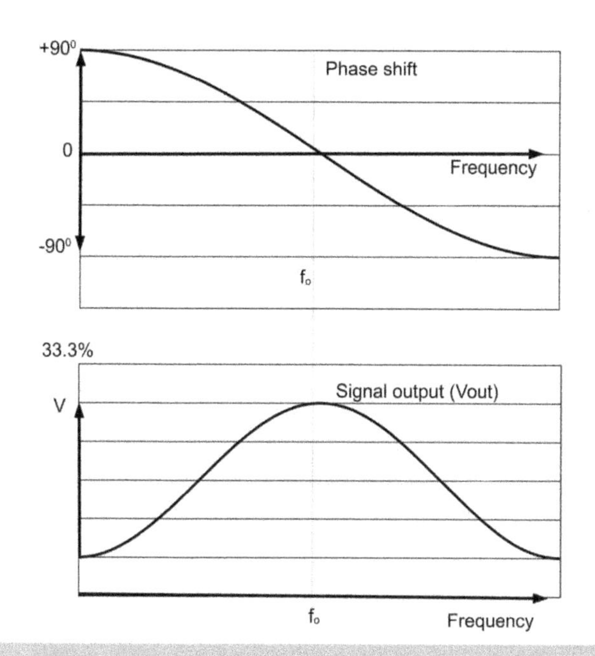

▲ **Figure 1.48** *Phase shift and gain loss of BP filter*

caused by the RC filtering components (33.33%), the op-amp requires a gain of 3. If the op-amp provides a gain of 3, the circuit will oscillate and a high quality, low distortion sine wave will be created (Figure 1.49).

If the gain of the op-amp is less than 3, oscillations will not sustain, if the gain is greater than 3, the oscillations will increase in amplitude until gross distortion occurs.

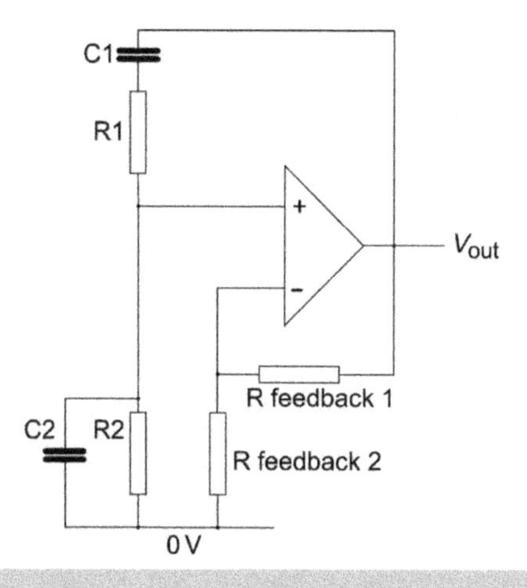

▲ **Figure 1.49** *Wien bridge oscillator*

Wien bridge oscillators, when adjusted correctly for a gain of 3, give the purest form of low frequency sine wave available.

The formula for gain of an op-amp connected for negative feedback, non-inverting is:

$$\text{Gain} = \frac{V_{OUT}}{V_{IN}} = \frac{\text{R1(feedback resistor 1)}}{\text{R2(feedback resistor 2)}} + 1$$

Therefore:

If $\text{Gain} = \frac{R1}{R2} + 1$ and Gain needs to be 3

$$3 = \frac{R1}{R2} + 1$$

$$3 - 1 = \frac{R1}{R2}$$

$$2 = \frac{R1}{R2}$$

So the ratio of the two feedback resistors must be 2 : 1, which will cause the op-amp to have a gain of 3, which will allow the circuit to just oscillate. Insufficient gain will cause the oscillations to decay; too much gain will cause the oscillations to increase until distortion occurs.

Transformers

Transformers are wound devices with many uses in electronics. We will discuss only two uses: voltage conversion (step-up, step-down) and impedance matching. Voltage conversion from one value to another is the most popular application for transformers.

Voltage step-up or step-down

A conventional transformer typically consists of two windings closely situated to each other. The windings are designated primary and secondary. When the primary winding has an AC voltage applied, a small current flows through the primary winding creating a magnetic field. This current is called the magnetising current. The magnetic field emanating from the primary winding cuts the secondary winding and induces an EMF (voltage) in the secondary winding. This induced voltage can be measured using a volt meter. The value of voltage induced depends upon the turn ratio of the transformer. The turn ratio is the number of turns of wire in the secondary winding divided by the number of turns in the primary winding.

$$\text{Ratio} = \frac{N_{\text{sec}}}{N_{\text{pri}}} \text{ Step-up/Step-down}$$

where N is the number of turns. We add the words step-up or step-down to ensure clarity. We usually quote turn ratios as being greater than one, for example, $\frac{500}{8} = 62.5$ written as 62.5 : 1 *step-down*. If we had the ratio the other way round, for example, $\frac{8}{500} = 0.016$ could be written as 0.016 : 1 which gives no instant feel to the size of the ratio. In this case we would write 62.5 : 1 *step-up*.

We should note that real world transformers are subject to losses. The windings are made of conductors such as copper, therefore they do have resistance and experience a voltage drop when current flows through a winding (copper losses or I^2R losses). The former, on which the transformer coils are wound, is affected by the magnetic field of the primary winding. A current is induced into the former (eddy current losses) further reducing the efficiency of the transformer. As more current is drawn from the secondary and primary windings of the transformer, I^2R loss increases, reducing the output voltage available. This loss of voltage while on load is referred to as transformer regulation and is indicated as a percentage of output voltage.

$$\text{Regulation} = \frac{V_{\text{offload}} - V_{\text{onload}}}{V_{\text{offload}}} \times 100\%,$$

For example, when the offload voltage = 100 V, the onload voltage = 95 V then the regulation $= \frac{100\text{ V} - 95\text{ V}}{100\text{ V}} \times 100\% = \underline{5\%}$

It should be noted that transformer regulation is measured and specified for a purely resistive load. A typical load consisting of diodes and capacitors will cause any transformer to suffer worse regulation due to the diodes and capacitors taking large current pulses during the voltage peak of each cycle.

If we disregard transformer losses, which we do unless instructed otherwise, we can use the formula: $V1 \times I1 = V2 \times I2$

As $P = V \times I$ leads us to:

Power input = power output.

Impedance matching

When a source (amplifier output etc.), AC or DC, is connected to a load, there will be an optimum impedance for the amplifier to get the maximum power transferred into the load. For example, an 8 Ω loudspeaker connected to an amplifier will only produce its maximum output if the amplifier has an 8 Ω output impedance. Figure 1.50 shows an amplifier with a range of output impedances connected to 8 Ω loudspeakers. Only 8–1C

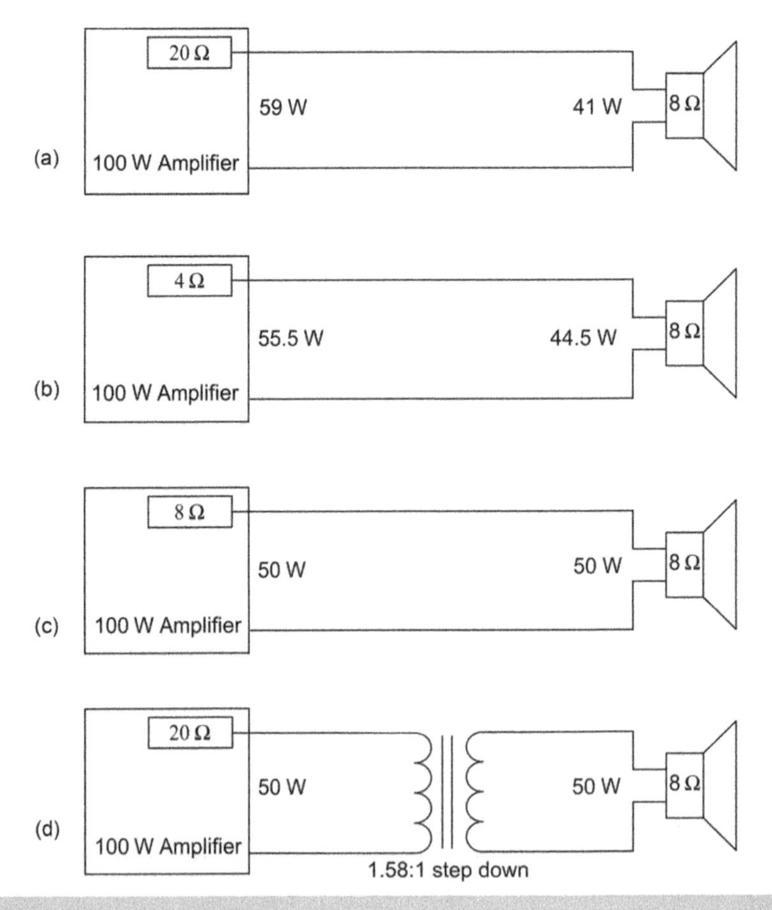

▲ **Figure 1.50** *Load matching using transformer*

and 8–1D offer a correct match. Figure 1.51 shows a graph of speaker impedance versus amplifier output impedance. The peak of the curve occurs when both are 8 Ω (matched).

The curve shows output power dissipated in a loudspeaker varying with impedance.

It can be seen that maximum power in the loudspeaker occurs when the loudspeaker impedance matches that of the amplifier. We usually do not have the luxury of changing the loudspeaker impedance or the amplifier impedance.

One option to overcome this mismatch is to use a matching transformer. In Figure 1.50, we have inserted a transformer to match the impedance of the amplifier to the loudspeaker. Using the formula:

$$\text{Ratio} = \sqrt{\frac{Z_s}{Z_p}}$$

where Z_p is the impedance of the amplifier and Z_s is the impedance of the loudspeaker.

Using the figures in Figure 1.50:

$$\text{Ratio} = \sqrt{\frac{8\Omega}{20\Omega}} = \sqrt{0.4} = 0.632 \quad \text{or} \quad 1.58:1 \textit{ step down.}$$

Therefore a transformer with a turn ratio of 1.58 : 1 will match an 8 Ω loudspeaker to an amplifier whose output impedance is 20 Ω. Any other transformer ratio will cause a loss of power in the loudspeaker and an increase in power dissipated in the amplifier.

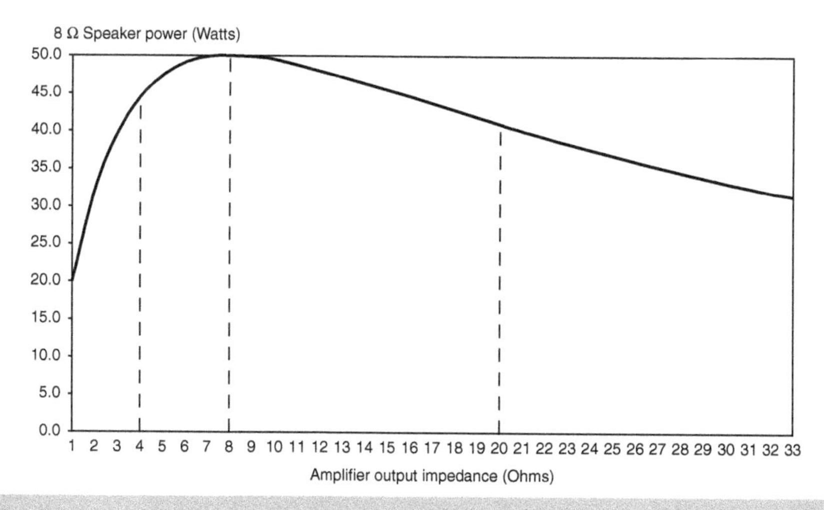

▲ **Figure 1.51** *Transformer matching chart*

To see the effect on the current within the output of the amplifier and load, we assume the amplifier is dissipating 100 W and therefore the best we can dissipate in the loudspeaker is 50 W.

In the primary circuit: $P = I^2R,\ I = \sqrt{\frac{P}{R}} = \sqrt{\frac{100}{20}} = 2.236\ \text{A}$

In the secondary circuit: $I = \sqrt{\frac{P}{R}} = \sqrt{\frac{100}{8}} = 3.536A$

If we multiply 2.236 A by the turns ratio of 1.58 we get 3.536 A, demonstrating matching power in the primary and secondary circuits when the correct turn ratio has been calculated using the impedance matching formula: $\text{Ratio} = \sqrt{\frac{Z_s}{Z_p}}$

An easy way to visualise impedance matching with transformers is to consider a 100 W lamp connected to a 200 V AC supply. The lamp will consume $P = VI$ watts, which means a current of $I = \frac{P}{V} = \frac{100\ \text{W}}{200\ \text{V}} = 0.5\ \text{A}$.

This shows us that the resistance of the lamp is

$$R = \frac{V}{I} = \frac{200\ \text{V}}{0.5\ \text{A}} = 400\ \Omega.$$

This can be seen in Figure 1.52(a).

If we were to connect the lamp to a 100 V supply (Figure 1.52(b)), the bulb will consume only one quarter of the power (25 W) and only give quarter the amount of light (assuming a linear transfer characteristic). To ensure the lamp consumes 100 W of power from a 100 V supply we could use a 'matching transformer'. In Figure 1.52(c) we can see a transformer inserted between the 100 V supply and the lamp. By use of a suitable step-up/step-down turn ratio we can increase the transformer secondary voltage to 200 V, which would allow the lamp to draw 0.5 A allowing 100 W to be dissipated. The turns ratio, in this example, can be seen by inspection to be 2 : 1 *step-up* (multiplying the input voltage by 2). The lamp does not 'know' that the supply is only 100 V, but it is drawing 100 W and the supply is giving 100 W. An ideal transformer with the correct turn ratio will transfer the same amount of power on the transformer input to the output, that is,

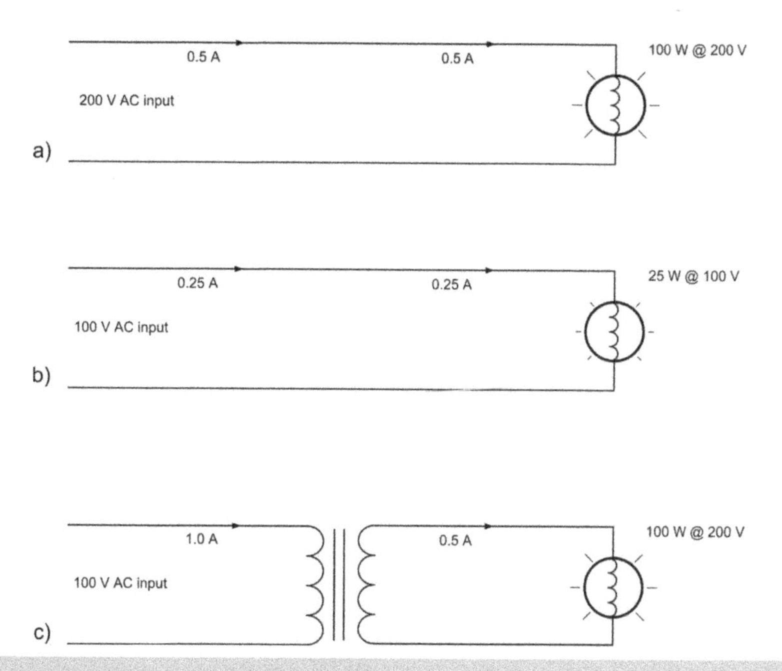

▲ **Figure 1.52** *Transformer matching example*

$$\text{Power in} = \text{Power out},$$

$$V_{in} \times I_{in} = V_{out} \times I_{out}\ \text{(Watts)}$$

In this type of application we would call this transformer a 2 : 1 *step-up*.

It is easy to get confused with step-up and step-down ratios when impedance matching. If you are going from high impedance to low impedance, for maximum power, the destination will require more current and less voltage, so you need a step-down ratio.

Amplifiers and Oscillators

Measurement of power and voltage gains/losses in decibels

It is used as a convenient way of describing the ratio of power in and power out of a system, and of describing the ratio of voltage in and voltage out of a system. Decibel is indicated by dB and it is a logarithmic ratio.

Decibels for power measurement

An amplifier usually increases the size of the signal at its input, and outputs this amplified signal.

The gain of the amplifier in power terms is $\frac{\text{Power Out}}{\text{Power In}}$.

For example, $\frac{10\,\text{W}}{1\,\text{W}}$ equals a gain of 10 (note – no units, this is just a ratio of two numbers).

To convert this gain to dB we use the following formula:

$$\text{dB} = 10 \times \log \frac{P_{out}}{P_{in}}$$

Using these figures

$$\text{dB} = 10 \times \log \frac{10}{1}$$

$$= 10 \times \log 10$$

$$= 10 \times 1$$

$$= \underline{10\,\text{dB}}$$

If you are given the gain of an amplifier as a value in dB and wish to convert it into the actual gain of the amplifier, use the following approach:

Assume that you have been told that the amplifier has a gain of 26 dB:

Using the dB power formula: $10 \times \log \frac{P_{out}}{P_{in}}$

$$10 \times \log \frac{P_{out}}{P_{in}} = 26\ \text{dB}$$

Divide both sides by 10 to get the 10 onto the RHS

$$\log\frac{P_{out}}{P_{in}} = \frac{26\text{ dB}}{10}$$

$$\log\frac{P_{out}}{P_{in}} = 2.6$$

Take the anti-log of both sides: $10^{2.6}$ = $\underline{398}$ (approx. 400)

$$\underline{\text{Gain}} = \underline{398}$$

What is the output of the 26 dB amplifier when 1 mW is applied?

$$0.001\text{W} \times 398 = \underline{0.398\text{ W}}$$

Decibels for voltage measurement

The gain of the amplifier in voltage terms is

$$\frac{\text{Voltage Out}}{\text{Voltage In}}$$

For example, $\frac{10V_{out}}{1V_{in}} = 10$ (note – no units, this is just a ratio of two numbers)

To convert this gain to dB we use the following formula:

$$\text{dB} = 20 \times \log\frac{V_{out}}{V_{in}}$$

Using the above example

$$20 \cdot \log\frac{10\text{V}}{1\text{V}} = 20 \cdot \log 10 = 20.1 = \underline{20\text{dB}}$$

If you are given the gain of an amplifier as a value in dB and wish to convert it into the actual gain of the amplifier, use the following approach:

Assume that you have been told that the amplifier has a gain of 46 dB, then

$$dB = 20 \times \log \frac{V_{out}}{V_{in}}$$

$$20 \cdot \log \frac{V_{out}}{V_{in}} = 46 \text{ dB}$$

Divide both sides by 20 to get the 20 onto the RHS

$$\log \frac{V_{out}}{V_{in}} = \frac{46}{20}$$

$$\log \frac{V_{out}}{V_{in}} = 2.3$$

Take the anti-log of both sides: $10^{2.3}$ = 199.52 (approx. 200).

What is the output of the amplifier when 1 μV is applied?

$$0.000001\text{V} \times 200 = 0.0002\text{V} = \underline{200\ \mu\text{V}}$$

Example

Figure 1.53 shows an amplifier with 40 dB of power gain. It has 75 Ω input and output impedance connected to a 75 Ω load. Apply 1 mW to the input. What is the power dissipated in the load resistor?

$$dB = 10 \times \log \frac{P_{out}}{P_{in}}$$

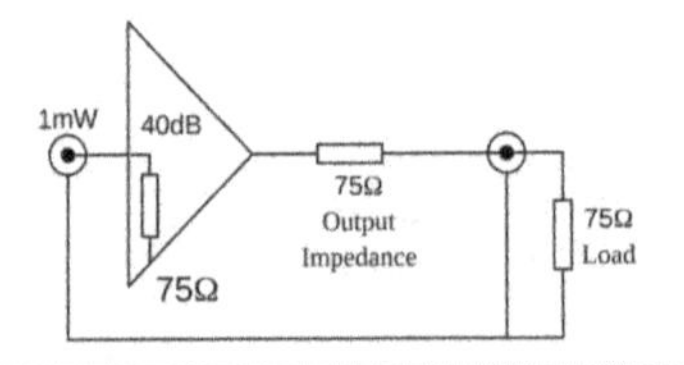

▲ **Figure 1.53** *Example with decibels*

Using the above figures:

$$40 = 10 \times \log\frac{P_{out}}{1\text{mW}}$$

$$\frac{40}{10} = \log\frac{P_{out}}{1\text{mW}}$$

$$4 = \log\frac{P_{out}}{1\text{mW}}$$

Take the anti-log of both sides:

$$10{,}000 = \frac{P_{out}}{1\text{mW}}$$

$$10{,}000 \times 1\text{mW} = P_{out} = \underline{10\,\text{W}}$$

Since output power is distributed between the amplifier output impedance and load impedance:

$$P_{load} = \frac{10\,\text{W}}{2} = \underline{5\,\text{W}}$$

Source of 1mW's output impedance is unknown so we assume 0Ω. Therefore 1mW arrives at the input.

Significant dB values

Every time we double the power output of an amplifier, we increase its output by 3 dB.

For example,

$$\text{dB} = 10 \times \log\frac{P_{out}}{P_{in}}$$

$\text{dB} = 10 \times \log\frac{2}{1}$ (say 2 W out for 1 W in)

$$\text{dB} = 10 \times \log 2 = 3\text{dB}$$

Doubling or halving the power gain of an amplifier is a change of 3 dB.

Every time we double the voltage output of an amplifier, we increase its output by 6 dB.

For example,

$$dB = 20 \times \log \frac{V_{out}}{V_{in}}$$

$$dB = 20 \times \log \frac{2}{1} \text{ (say 2 V out for 1 V in)}$$

$$dB = 20 \times \log 2 = 6 \text{ dB}$$

Doubling or halving the voltage gain of an amplifier is a change of 6 dB.

Attenuation

Attenuation can be viewed as the opposite of gain. Attenuation reduces the level of the signal passing through an attenuator. Attenuation is always measured in decibels. Attenuation can be applied to both power and voltage.

The usual dB formula apply:

$$dB = 10 \times \log \frac{P_{out}}{P_{in}}$$

$$dB = 20 \times \log \frac{V_{out}}{V_{in}}$$

Power attenuation

P_{out} will always be smaller than P_{in}. This will give the dB value a negative sign.

For an attenuator with 2 W applied to its input, giving 1 W output:

$$dB = 10 \times \log \frac{P_{out}}{P_{in}}$$

$$dB = 10 \times \log \frac{1}{2}$$

$$dB = 10 \times \log 0.5 = -3 \text{dB}$$

Voltage attenuation

V_{out} will always be smaller than V_{in}. This will give the dB value a negative sign. For an attenuator with 2 V applied to its input, giving 1 V output:

$$dB = 20 \times \log\frac{V_{out}}{V_{in}}$$

$$dB = 20 \times \log\frac{1}{2}$$

$$dB = 20 \times \log 0.5 = \underline{-6\text{ dB}}$$

Decibels are used to easily refer to large changes in large numbers. For example, to say that a signal has increased 50 dB is much easier than to say the output has increased 100,000 times. Whether the input power is increased from 1 to 100 W or from 1,000 to 100,000 W, the amount of increase is still 20 dB. In Table 1.4, we can see dB to power ratio conversions.

Table 1.4 *Decibel power ratios*

dB	Power Ratio	
1 =	1.3	
3 =	2.0	
5 =	3.2	
6 =	4.0	
7 =	5.0	
10 =	10	$=10^1$
20 =	100	$=10^2$
30 =	1,000	$=10^3$
40 =	10,000	$=10^4$
50 =	100,000	$=10^5$
60 =	1,000,000	$=10^6$
70 =	10,000,000	$=10^7$
100 =	10^{10}	
110 =	10^{11}	
140 =	10^{14}	

From Table 9.1 we can see that 3 dB has a power ratio of 2 (doubling or halving power); dBs can be added together for amplifier stages that are connected in series.

For example, if two amplifiers are connected in series, with gain being 2 and 5 respectively. The overall gain would be $2 \times 5 = 10$. If we used dBs to indicate the gain, using the formula $dB = 10 \times \log \frac{V_{out}}{V_{in}}$ or referring to Table 1.4 we would have gains of 3 dB and 7 dB respectively. To calculate the gain of an amplifier in dBs we just add the dBs: 3 dB + 7 dB = 10 dB.

As shown in Table 1.5, a gain of 10 is 10 dB; adding dBs is the same as multiplying the pure gain.

Table 1.5 *Decibels and gain*

Gain 1		**Gain 2**	**Total Gain**
3 dB	+	7 dB	=10 dB
2	x	5	=10

Signal-to-noise ratio

Signal-to-noise ratio (SNR) is defined as the ratio of the signal power to the noise power.

$$SNR = \frac{\text{Signal power}}{\text{Noise power}}$$

It can of course be defined using decibels:

$$SNR = 10 \cdot \log \frac{\text{Signal power}}{\text{Noise power}} \text{dB}$$

SNR or S/N is important in electronic amplifiers as a measure of both how much noise an amplifier adds to the signal as it is being amplified, and the minimum amplitude of signal that can be realistically received by a receiver. For example, a low noise receiver has a SNR of 40 dB. This means that (refer to Table 1.4) 40 dB is a power ratio of 10,000 to 1. For every 10,000 units of signal there will be 1 unit of noise added by the receiver as it amplifies the signal. The signal is wanted, noise is unwanted. Electronic circuitry, resistors, transistors, etc., all produce noise when an electrical current is flowing. The aim of electronic designers is to minimise the amount of noise, while enhancing the quantity of signal. (High signal and low noise gives a large SNR.)

Figure 1.54 shows a spectrum analyser display showing a 50 dB, 90 MHz signal with noise barely measurable. The noise is the line at the bottom of Figure 1.54. The SNR is 50 dB – 0 dB = 50 dB. Figure 1.55 shows a plot with a much lower (worse) SNR, the noise is approximately 13 dB, the wanted signal at 50 dB, giving an SNR of 50 – 13 = 37

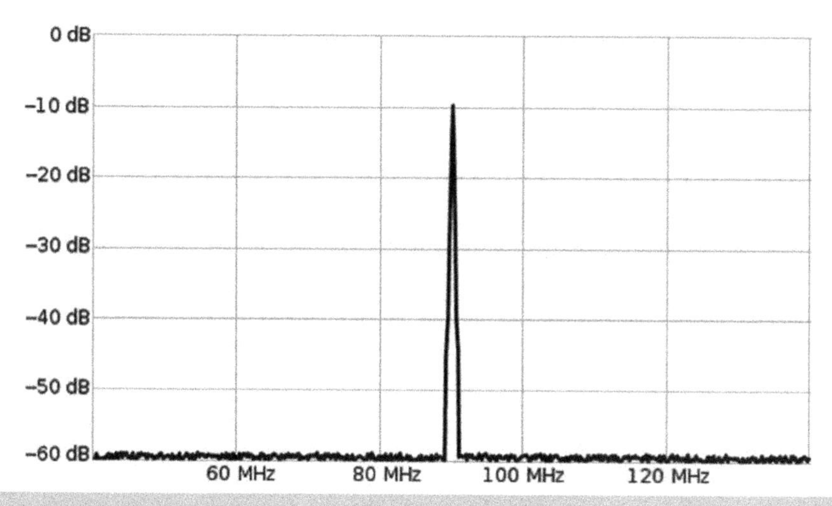

▲ **Figure 1.54** *Spectrum analyser – good S/R*

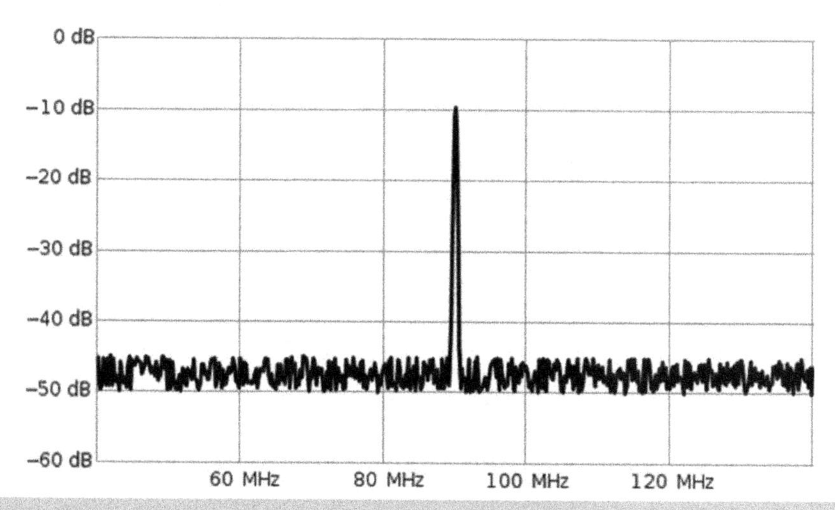

▲ **Figure 1.55** *Spectrum analyser – poor S/R*

dB. Remember that dBs are logarithmic so subtracting dBs is the same as dividing the original power measurements. SNRs can be measured in a variety of ways, depending upon bandwidth of the system and how the noise level is handled.

Transistor as a switch

Up to now we have considered the transistor to be an analogue device: we can vary the base current and the collector varies greatly. However, a transistor can also be used as a switch. In Figure 1.56 we can see a transistor connected with a 1 kΩ resistor from its collector to the upper power rail. The transistor base is connected via a 10 kΩ resistor to a potentiometer, which is connected between the power rails. The potentiometer can be adjusted to place any voltage between 0 and 10 V onto the left hand end of the 10 kΩ resistor. With 0 V on the 10 kΩ resistor, that is, zero current flowing to the base of the transistor, the transistor is Off/Not conducting/Open circuit.

As the potentiometer is adjusted to give a higher voltage on the left hand side of R2, the amount of base current increases. The collector current increases, tracking the base current multiplied by the transistor gain.

In Figure 1.57 we can see a switch SW1 which can force current into the base of the transistor. When the switch is not actuated, the base of the transistor is connected to 0 V and is off (not conducting). Whenever the switch is actuated, the base of the transistor is connected, via the 10 kΩ resistor to 10 V. This causes a current to flow through the

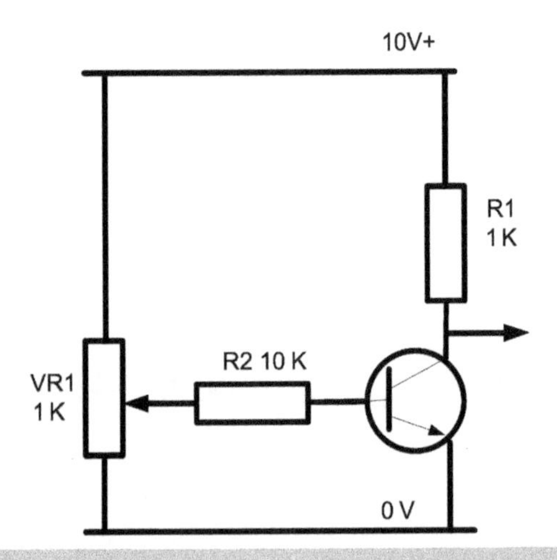

▲ **Figure 1.56** *Simple transistor circuit*

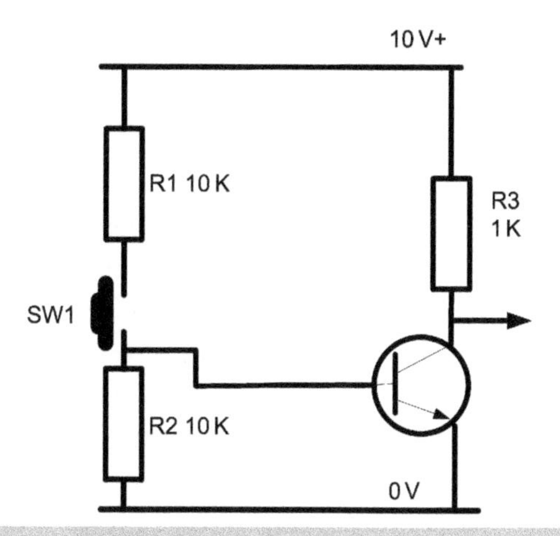

▲ **Figure 1.57** *Simple transistor switch circuit*

transistor base/emitter and through R1 (10kΩ) resistor. The current is limited to less than 1mA due to the value of this resistor:

$$I = \frac{V}{R}$$

$$I = \frac{10 - 0.7}{10000}$$

Base current = 930 μA.

Actuating (closing) the switch causes current to flow through the resistor and B–E junction, forward biasing the transistor base/emitter junction, turning the transistor on, causing current to flow through the emitter collector path. For silicon transistors, due to the semiconductor junction effect, a voltage of approximately 0.7 V can be measured between the base and emitter terminals V_{BE} when sufficient current is forced through the base/emitter junction. With base/emitter junction carrying current, current flows from the emitter to collector terminal. The amount of current flowing from emitter to collector depends upon the amount of base current and the gain of the transistor. With the transistor carrying the maximum amount of current (fully saturated), the voltage between the emitter and collector terminals is typically 0.2 V for a silicon transistor. When the base current drops to 0 A due to the switch being open, the collector current drops to 0 A and the collector to emitter V_{CE} voltage rises to the maximum available.

In this manner, we can say that a transistor can act as a switch because we can turn the transistor On and Off by use of a small current fed through the base.

Linear amplifier

Figure 1.58 shows the relationship between base current I_B and collector current I_C . If the base current of the transistor was adjusted to be mid-way between zero and the maximum value required to allow the chosen value of collector current to flow, then any small change in base current caused by the application of a small AC signal would cause a linear change in collector current. If the variation of base current I_B remained on the linear part of the curve in Figure 1.58 then the relationship between I_B and I_C currents would remain linear. This is called linear operation.

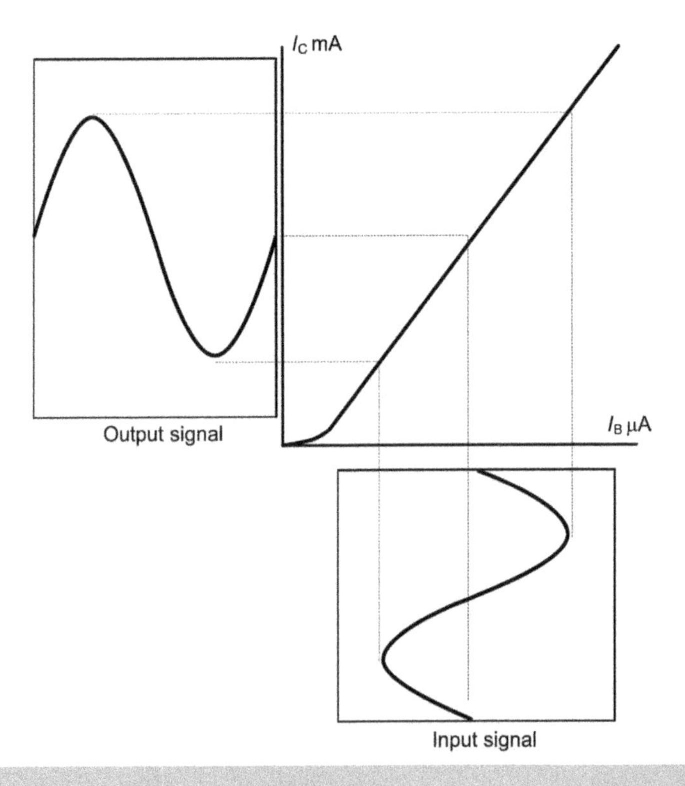

▲ **Figure 1.58** *Amplifier with linear input and output*

Linear operation

The static amount of current could be reduced causing I_C to no longer follow I_B, this would result in severe non-linearity. This non-linear operation is sometimes used in specialised radio transmitters to increase efficiency. But if the amplitude characteristics of the waveform need to be retained, then linear amplification is necessary. Single sideband transmitters use linear amplifiers to ensure their transmitted signal does not become distorted.

Power amplifier

Power amplifiers are devices that are designed to deliver power into a load, for example, audio amplifiers deliver power into a loudspeaker or radio transmitters delivers power into an antenna.

Power amplifier final stages (where power is coupled to a load) differ from a standard amplifier to enable the amplifiers maximum power at maximum efficiency to be delivered to the load.

Constant current generator

Constant current generators can be developed which give a constant current, irrespective of the load resistance or applied voltage. Constant current generators can be fed with a variable voltage, the current output, tracking the voltage supplied; these are called voltage to current converters.

In Figure 1.59 an op-amp is shown driving current down a line towards R_{load}. The 1–5 V signal applied to the non-inverting input. The value of the 250 Ω resistor ensures that with 5 V applied to the op-amp, a current of 20 mA will flow through the circuit and with 1 V applied, 4 mA will flow through the circuit. The op-amp is supplying the current to the load, so it will need to have good current capabilities and sufficient voltage supply. If the op-amp is under-powered and cannot drive the required current, then

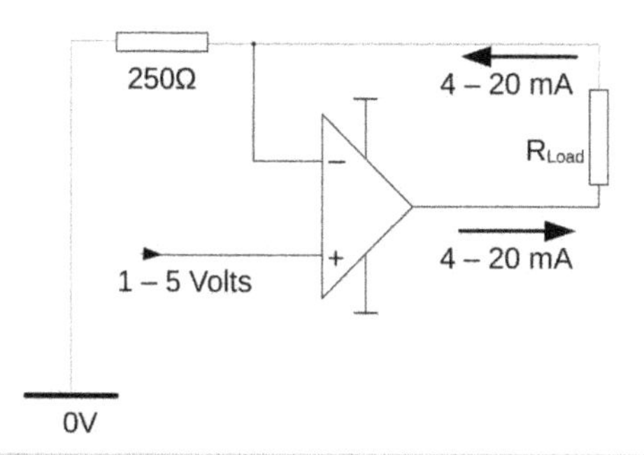

▲ **Figure 1.59** *Voltage to constant current converter (constant I)*

a transistor can be connected to the output of the op-amp to increase its capability. Constant current generators form part of a 4–20 mA loop control system.

The control signal to open or close a valve or the signal from a sensor would vary the voltage input to the circuit in figure 1.59. The variable current output can then be sent over long cables without degradation.

Another common use it to supply a constant current to a 4 wire PT100 sensor. By feeding the sensor with a constant current, cable length will have no effect of the temperature reading.

LC and RC sinusoidal oscillators

Resistor capacitor sinusoidal oscillators are usually implemented by means of a Wien bridge circuit. Inductor–capacitor (LC) oscillators are implemented using an inductor (L) and capacitor (C) in a variety of configurations.

An LC combination can, when shocked, oscillate in the same manner as a playground swing. For example, if an electrical charge was caused to flow into a capacitor and that capacitor was connected in parallel to an inductor, when the capacitor discharged, the electrons would tend to flow from the capacitor into the inductor, and the current flowing into the inductor would tend to generate magnetic lines of flux. When the capacitor has discharged sufficiently, the inductor's magnetic field will start to collapse, causing a current to flow back into the capacitor. This 'backwards and forwards' motion will continue until all the power has been dissipated by the resistance in the circuit. Figure 1.60(a) shows a parallel LC circuit where the current/voltage oscillates between the parallel LC components. Figure 1.60(b) shows a series LC circuit where the current/ voltage oscillates between the two LC components.

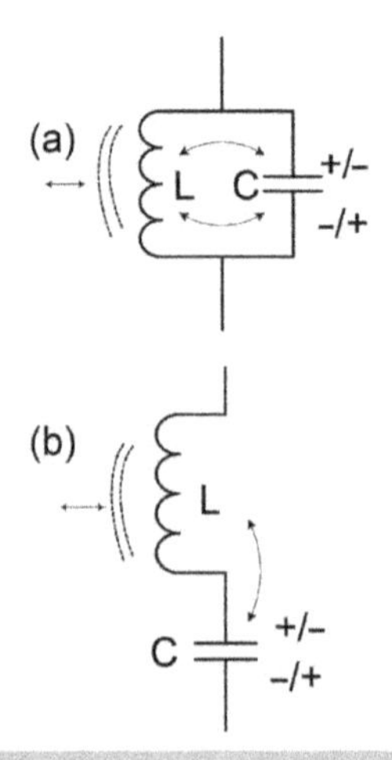

▲ **Figure 1.60** *LC circuit in series and parallel*

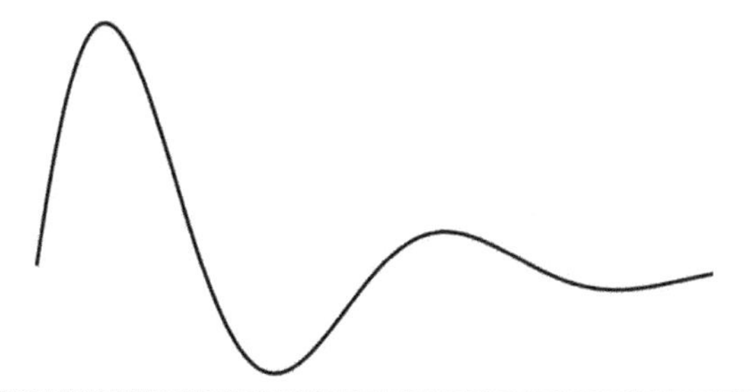

▲ **Figure 1.61** *Decaying sine wave*

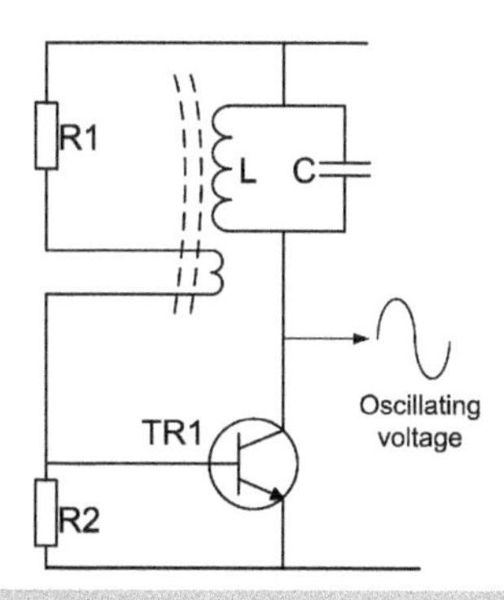

▲ **Figure 1.62** *Transistor LC oscillator*

In Figure 1.61, we can see a decaying sine wave due to the resistance within the L and C, dissipating power as the current oscillates from L to C and back. In Figure 1.62, we can see that the coil L has a small secondary winding. This secondary winding is connected to the base of the transistor TR1. The signal from the secondary winding is in phase with the oscillation occurring within LC. This adds to the oscillation.

The gain of the transistor overcomes the losses within the LC components. By choosing a transistor of sufficient gain and a coil with a suitable turns ratio, sufficient feedback will be provided to sustain the oscillation indefinitely.

The frequency of oscillation depends upon the value of L and C:

$$f_o = \frac{1}{2\pi\sqrt{LC}} \text{Hz}$$

Square wave generators

Square waves are very useful in digital circuits and are usually generated by specialist devices. One of the classic devices to generate a square wave is a 555 timer integrated circuit. The 555 can be connected and used in a variety of modes depending on what

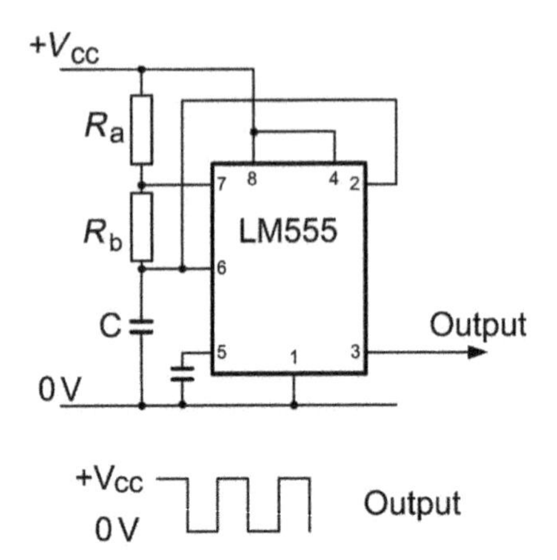

▲ **Figure 1.63** *555 astable oscillator (free running)*

operation is required. In Figure 1.63, we can see a 555 connected in an astable mode. This means that the output is continuously changing from a high level to a low level, at a frequency determined by the values of $R_a R_b$ and C. The mark space ratio of the output is determined by the ratio of the two resistors R_a and R_b .

Another use of the 555 timer device is as a monostable. This is seen in Figure 1.64 where a pulse applied to the input of a 555 triggers a longer, controlled width pulse from the device output.

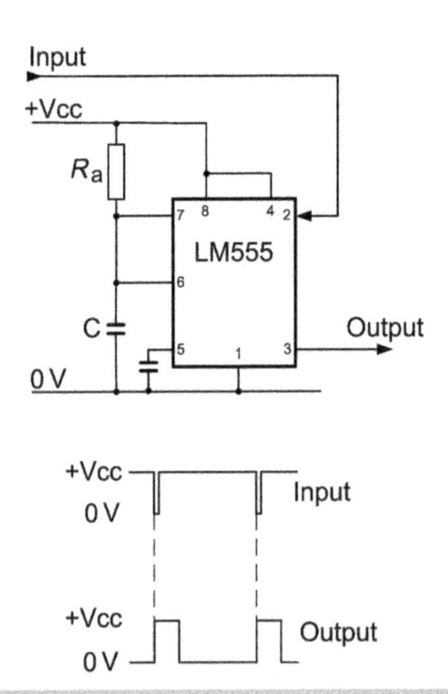

▲ **Figure 1.64** *555 monostable (requires a trigger pulse)*

To calculate the frequency of oscillation or the width of the output pulse, many manufacturers of 555 devices provide both formulas and nomograms. In Figure 1.65 a nomogram can be seen which allows you to choose the value of the resistor/capacitor combination to achieve your desired output frequency.

In Figure 1.66 a nomogram can be seen which allows you to choose the value of the resistor/capacitor combination to achieve your desired output pulse width or delay time before the device output pin returns to its idle state.

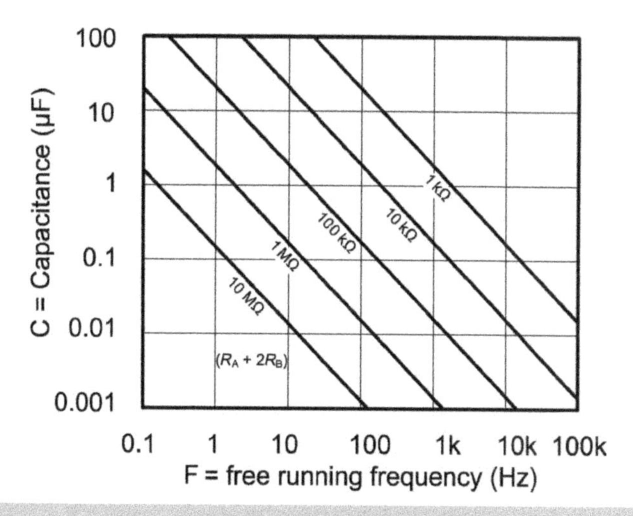

▲ **Figure 1.65** *555 astable nomogram. © Texas Instruments www.ti.com*

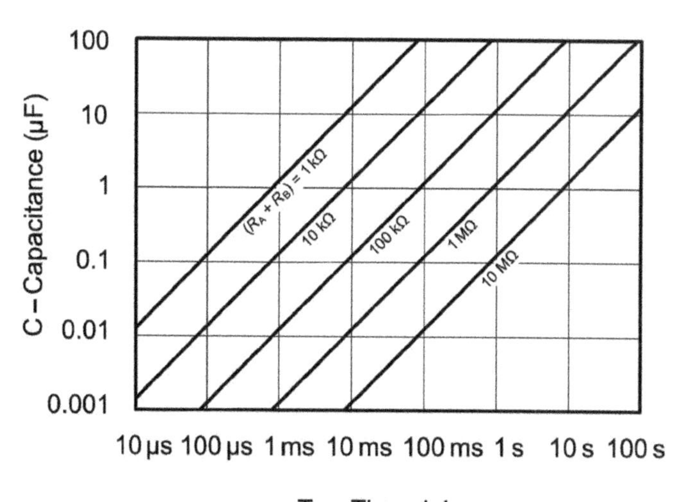

▲ **Figure 1.66** *555 monostable nomogram. © Texas Instruments www.ti.com*

The formula for the free running frequency of a LM555 device manufactured by a National Semiconductor is:

$$f_o = \frac{1.44}{(R_a + 2R_b)C} \text{Hz}$$

When in monostable configuration, the width of the output pulse or delay time from pulse start to pulse end is given by the formula:

$$t = 1.1 R_a C \text{ seconds}$$

Power Supplies

Electronic equipment requires power supplies to provide a range of voltages and currents to enable the equipment to work. For small equipment typical voltages could be 5, 12 and 24 V. When these voltages are not available, we need a power supply circuit to convert AC mains voltage into these low voltages. As most mains supplies are AC, it is typical to use a transformer to convert the 230 V AC to a lower voltage such as 24 V AC (Figure 1.67). We can then rectify (convert from AC to DC) and deliver a controlled DC voltage.

To rectify (convert from AC to DC) we need a diode which allows voltage and current to pass through it in only one direction.

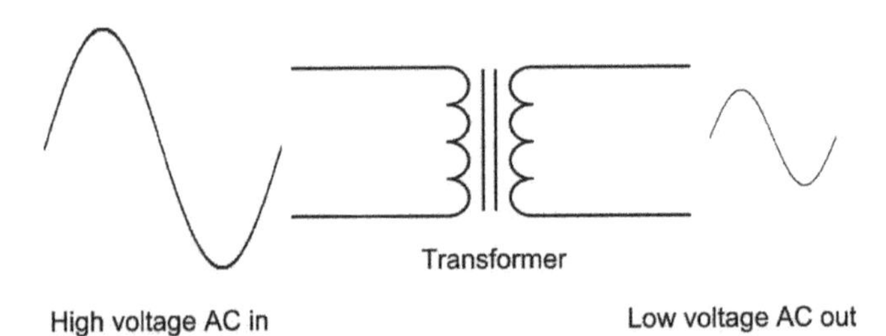

▲ **Figure 1.67** *Transformer action*

Half wave, full wave, bridge rectifiers

A variety of diode configurations can be used to convert AC into DC. The simplest is a single diode connected as shown in Figure 1.68. An AC mains voltage is fed in on the left, into the primary winding, and a much reduced AC voltage appears on the right, from the secondary winding.

This AC voltage is passed to a diode which only allows voltage and current to pass one way. This effectively cuts the AC waveform in half (hence the name half-wave rectifier). The diode output voltage is shown in Figure 1.69.

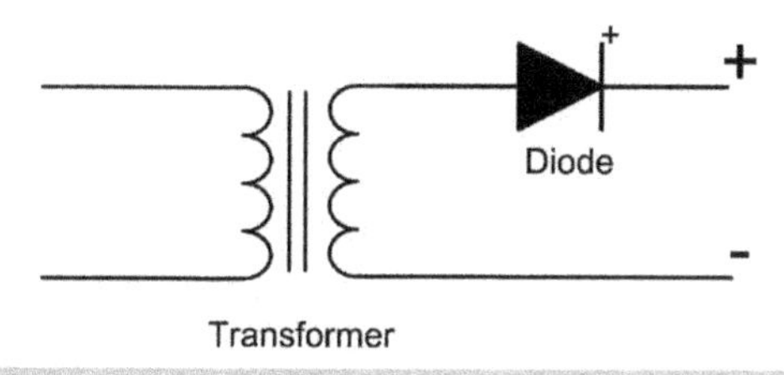

▲ **Figure 1.68** *Rectifier action*

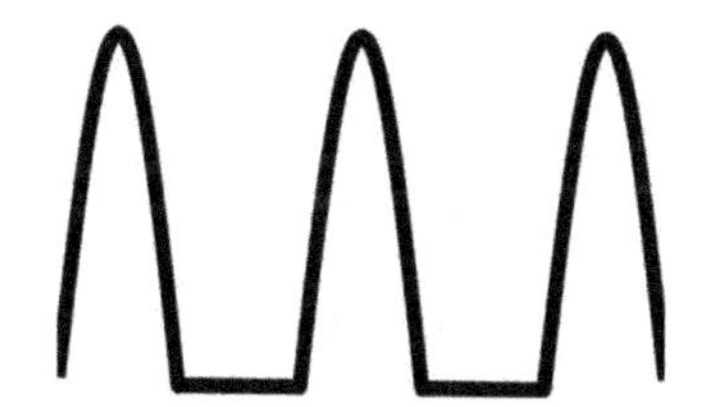

▲ **Figure 1.69** *Half wave diode output*

Diode ratings

Diodes, when used as power rectifiers, have two major parameters to consider: peak inverse voltage and maximum forward current. When diodes are used in a power supply circuit to convert AC to DC, we need to consider the maximum voltage that will be applied to the diode. This is the voltage coming from the step-down mains transformer. The diode needs to be able to withstand the voltage from the secondary winding. This is called the peak inverse voltage. The maximum forward current is the amount of current the diode can conduct without melting.

Smoothing and regulation

Figure 1.69 shows a very peaky output voltage with the average voltage level quite low. On inspection, we can see that half of the input waveform has been removed. This is very wasteful. We have generated a complete AC voltage, only to have our simple diode rectifier remove one half. A more efficient way to proceed would be to create a diode rectifier circuit which allowed both halves of the AC waveform (the +ve and the –ve half) to be converted into DC. This is what we do with a full-wave rectifier.

Figure 1.70 shows the diode bridge added to the circuit. We can see that the output of the diode bridge consists of both half-waves, giving a much greater average voltage. We can reduce the voltage variations (ripple) if we add a capacitor across the output terminals. The capacitor is charged up to the peak voltage from the bridge rectifier, and when the voltage from the bridge reduces, the capacitor takes over as the source of voltage/current for the circuit. In this way, the capacitor helps to maintain the output voltage of this circuit. The larger the value of capacitor (in Farads) the smaller will be the resultant ripple. Capacitors used in this way need to be carefully specified to ensure that they can accept large peak currents rushing into the capacitor. We can see a capacitor added to the circuit and its resultant smaller ripple in Figure 1.71.

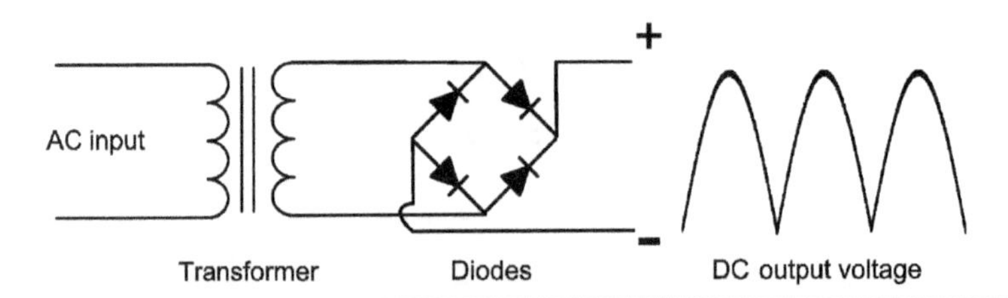

▲ **Figure 1.70** *Full wave diode output*

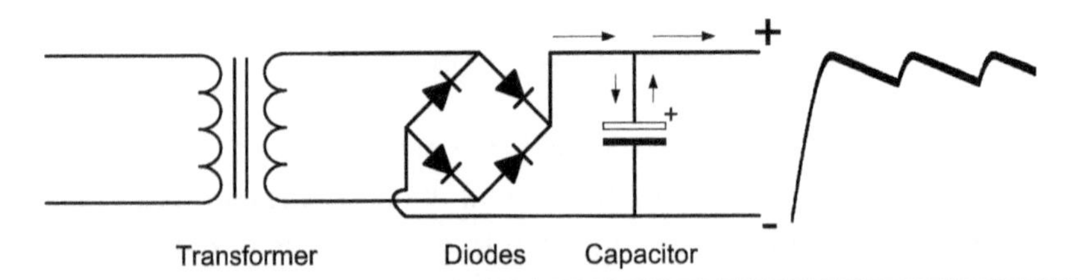

▲ **Figure 1.71** *Full wave diode output – smoothed with capacitor*

Switch mode power supplies

Switch mode power supplies are very efficient at changing voltage levels and current capacities.

Switch mode supplies (SMS) utilise the energy stored in an inductor to give increased voltage outputs and use a variable pulse width modulation system to control the output voltage. Figure 1.72a shows a simplified circuit of an SMS. Tr1 is fed with a stream of pulses from the 555 astable causing Tr1 to switch on and off rapidly, typically at 100 kHz. Each time Tr1 is turned off, the collapse of current through Tr1 and L1 causes a high voltage pulse to be developed across L1 (the same type of pulse that relay coils generate). This pulse is rectified by D1, and smoothed by C1 and C2 into smooth DC. As the transistor is either fully on or fully off, it dissipates little power, whereas a linear PSU is constantly consuming power. Due to the small amount of power dissipated by the transistor, switching power supplies are among the most efficient available. In a complete circuit, the voltage output would be fed back to the 555 function, when more current was needed by the load, causing the SMS voltage output to dip, and the mark space ratio of the 555 would change to give more energy into L1 and increase the voltage rectified by D1.

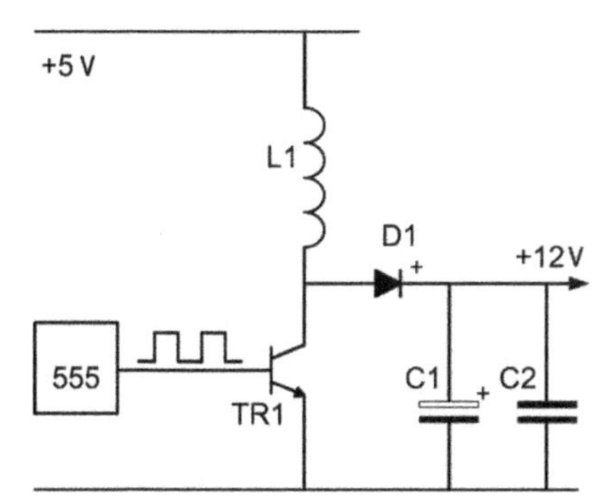

▲ **Figure 1.72a** *Switching power supply (simplified)*

In figure 1.72b we can see RV1 providing voltage feedback to the system. If load in increased and the output voltage dips, the circuit will change the pulse width to bring the voltage back to the desired value. In a similar way, R1, being in series with the output current monitors the current flow. If the current becomes to high, the increasing voltage dropped across R1, fed back to the circuit will reduce the pulse width to drop the output voltage.

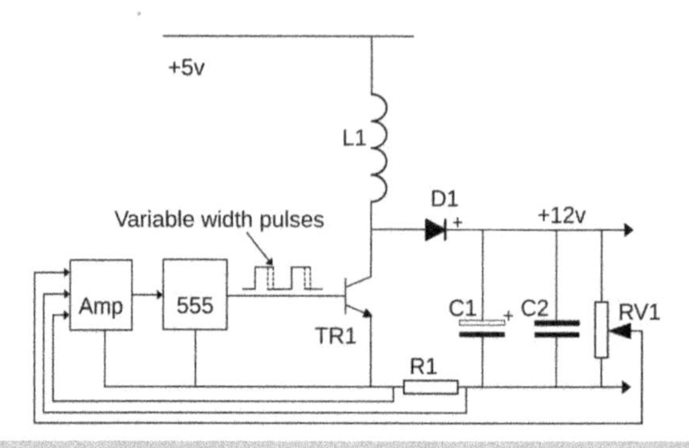

▲ **Figure 1.72b** *Switching PSU with regulation and current limiting*

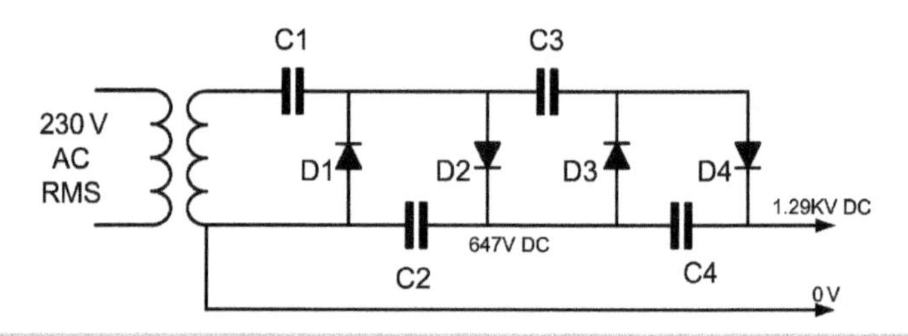

▲ **Figure 1.73** *Voltage multiplier*

Voltage multipliers

Figure 1.73 shows a typical voltage multiplier circuit. An AC voltage of 230 V RMS is applied to the primary of the transformer. The transformer turn ratio is 1 : 1. The secondary voltage is therefore 230 V RMS. To convert an RMS voltage value to peak voltage we multiply by 1.414. The peak voltage is 230 × 1.414 = 325.22 V. The peak to peak voltage value is the peak voltage multiplied by 2, 325.22 V × 2 = 650.44 V. This peak to peak waveform is fed from the secondary winding to C1 and D1. D1 clamps this peak to peak voltage to 0V, but C1 allows the whole voltage change to pass through. The voltage at the junction of C1 and D1 is a sine wave, from 0 to 650 V. This sine wave is coupled by C2 to D2, and D2 rectifies the sine wave to 647 V DC. This voltage could be used if required. The next stage of the multiplier (C3, C4, D3 and D4) rectifies and adds the 650 V sine wave from the junction of C1 and D to the newly created 647 V DC, giving 1.3 kV and the junction of C4 and D4. This process of adding stages to the multiplier can continue many times. In this example each additional stage increases the output voltage by 647 V DC. If more than a small current is used by a load connected to this voltage multiplier, the output voltage will drop and dramatically increase the size of any ripple voltage.

Digital Devices and Systems

Logic gates: AND, OR, NOT, NAND and NOR

Digital circuitry relies upon Boolean logic. Boolean logic has two states: true and false, conveniently represented by 1 and 0. Boolean logic allows arithmetic to be undertaken using these values of 1 and 0. Under Boolean rules:

0 AND 0 = 0

0 AND 1 = 0

1 AND 0 = 0

1 AND 1 = 1

0 OR 0 = 0

0 OR 1 = 1

1 OR 0 = 1

1 OR 1 = 1

NOT 0 = 1

NOT 1 = 0

The three functions AND, OR and NOT can be used to build other functions such as NAND and NOR. To see simple Boolean logic in operation let's see how we can turn a light on with two switches.

In Figure 1.74(a) we see two push buttons in series connected to a light bulb and a battery. We say these push buttons are in series with each other, the Boolean term is AND, we need button 1 AND button 2 pressed to make the light glow. If we could achieve this with electronics, we would see the circuit and symbol as in Figure 1.74(b),

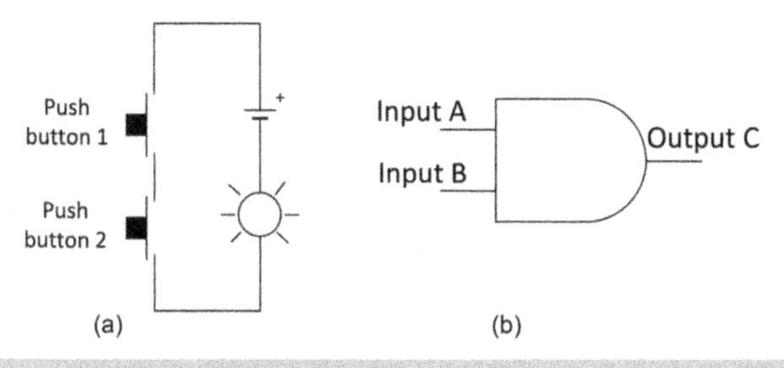

▲ **Figure 1.74** *AND logic*

Table 1.6 *Truth table for a 2 input AND gate*

Input A	Input B	Output C
0	0	0
0	1	0
1	0	0
1	1	1

Table 1.7 *Truth table for a 2 input OR gate*

Input A	Input B	Output C
0	0	0
0	1	1
1	0	1
1	1	1

an AND gate with 2 inputs and 1 output. To represent an electronic gate Boolean function we typically use a truth table (Table 1.6).

The OR function can similarly be represented by switches, gates or a table (Table 1.7).

In Figure 1.75(a) we see two push buttons in parallel connected to a light bulb and a battery. We say these push buttons are in parallel with each other, the Boolean term is OR, we need button 1 OR button 2 pressed to make the light glow. If we could achieve this with electronics, we would see the circuit and symbol as in Figure 1.75(b), an OR gate with 2 inputs and 1 output.

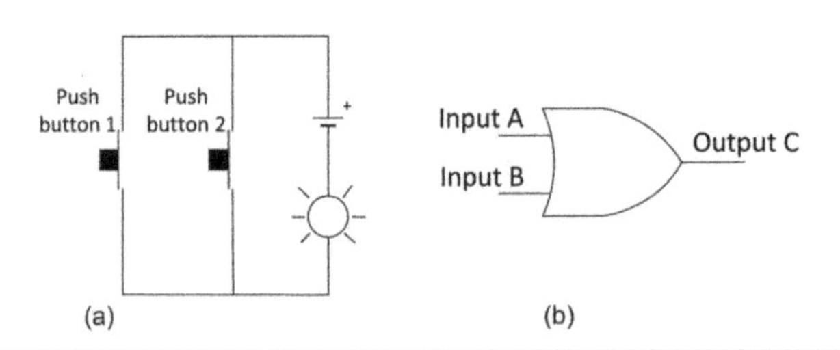

▲ **Figure 1.75** *OR logic*

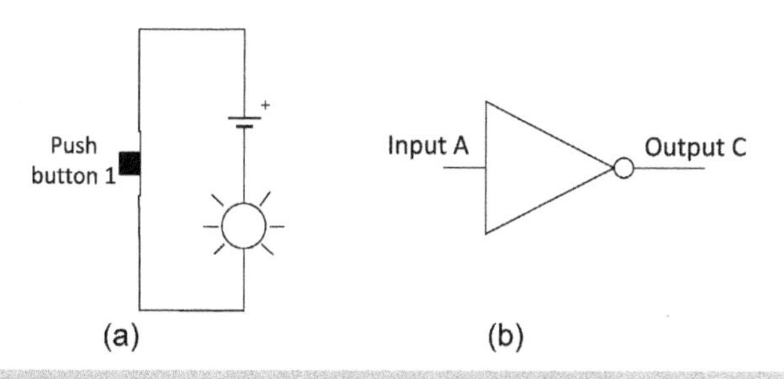

▲ **Figure 1.76** *NOT logic*

Table 1.8 *Truth table for a NOT gate*

Input A	Output C
0	1
1	0

In Figure 1.76(a) we see 1 normally closed push button connected to a light bulb and a battery. We say this push button inverts the normal action of a normally open switch. To turn the light off we need to push the button, we need button 1 NOT pressed to make the light glow. If we could achieve this with electronics, we would see the circuit and symbol as in Figure 1.76(b), a NOT gate with 1 input and 1 output. We also call a NOT function an inverter as we invert the logic level (Table 1.8).

We can connect gates together to create large, more complex functions. Before we cover combinational circuits let us look at how we connect logic gates together.

Logic levels and families

So far we have discussed logic gates without discussing how they work. Logic gates are small digital electronic devices which are usually made out of silicon and consist of a collection of transistors, diodes and resistors. Single transistors have specific voltage and current requirements to allow them to function in useful ways. When transistors are built into a single 'Integrated Circuit' (IC), the voltage and current values that we used for individual transistors no longer apply. Silicon digital ICs implementing logic functions with transistors that are NPN or PNP types (bi-polar) require a voltage supply of 5.0 V DC. Logic levels of 1 and 0 are required to map to real voltages that each gate input and output understand. The original family of digital ICs were named Transistor, Transistor Logic (TTL). TTL device inputs accept voltage inputs for logic 0 of <0.8 V, and logic 1 of >2.0 V. TTL logic output thresholds are logic 0 <0.4 V and logic 1 >2.7 V.

These voltages and how they relate to inputs and outputs can be best seen in Figure 1.77, where two gates are connected together.

In Figure 1.77 two gates are connected together from IC1–3 to IC2–1. Logic levels leaving IC1–3 are in reality voltage levels that have to match the expectation of the join lines together receiving gate IC2.1. If TTL logic 1 has to be 2.7 V or greater, then this logic 1 will always be received correctly by IC2.1, as it is looking to receive a logic 1 if the voltage level exceeds 2.0 V. Similarly for logic 0, the transmitting gate IC1.3 outputs a voltage level of less than 0.4 V but the receiving gate will tolerate any voltage below 0.8 V as logic 0. Output voltage levels are designed to exceed the requirements of any connected inputs to ensure correct logic levels are communicated from device to device. Each input connected to an output consumes a small amount of current (40 μA). Each output can provide 400 μA, so each output can drive (or be connected to) 10 inputs. This is called 'fan-out', for example, an output with a fan-out of 10 can be connected to up to 10 inputs and still achieve the required voltage levels. Similarly, inputs have a 'fan-in'. Inputs require a certain amount of current to operate correctly, in this case we are assuming 40 μA, but some inputs on some devices may have a current requirement of 40 μA or more. This is indicated by the fan-in figure, the lower the fan-in, the fewer inputs can be connected to the same output.

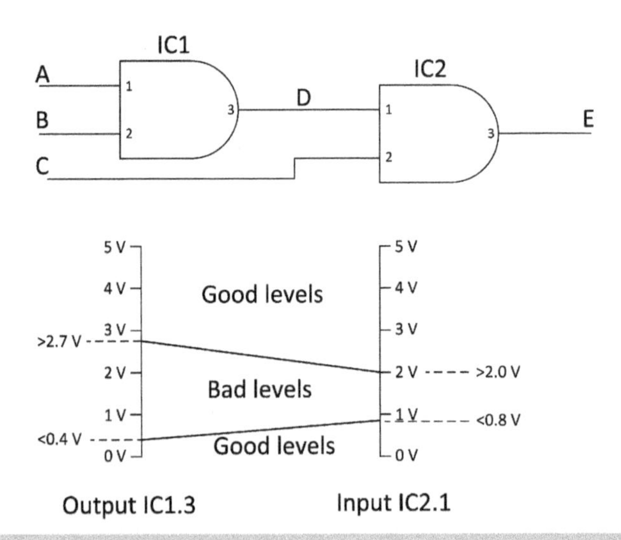

▲ **Figure 1.77** *Threshold levels (input and output)*

Combinational circuits

We have shown above that it is possible to connect one gate to another. Now we explore this further. In Figure 1.78 we can see a more complex set of gates.

If each gate performs a simple logic function and outputs change immediately when an input is changed, we describe this type of circuit as a 'combinational circuit' or 'combinatorial circuit'. The output logic level depends upon the combined logic functions of each gate, connections and input logic levels. In Figure 1.78 we have three gates, IC1 and IC2 are AND gates, IC3 is an OR gate. Inputs A, B, C and D connect to the outside world along with the circuit output G. Intermediate connections are labelled E and F. The truth table for IC1 can be seen in Table 1.9, columns A, B and E; IC2 is shown in columns C, D and F. You can see the similarity with the AND truth table shown earlier. The OR gate truth table for IC3 can be identified in columns E, F and G. To calculate what state output F should be for any combination of inputs, we methodically fill in Table 1.9 starting with IC1. The value for column E is the AND of A and B, so only the last 4 entries in column E are set to 1, the other entries are set to 0. The value for column F depends upon the AND of C and D, so there are 4 occurrences when F is set to 1. Output F depends upon the OR of values E and F. There are 7 occurrences of 1 in column G. For any combination of binary levels on input pins A, B, C and D we can determine what output G will be.

Table 1.9 shows us the logic levels when the circuit is working as expected. Logic levels in columns A to D is a simple binary pattern from 0000 to 1111. Columns D and E are intermediate signals, and column G is the final output signal.

How do we approach finding a faulty digital component in a non-working version of this circuit?

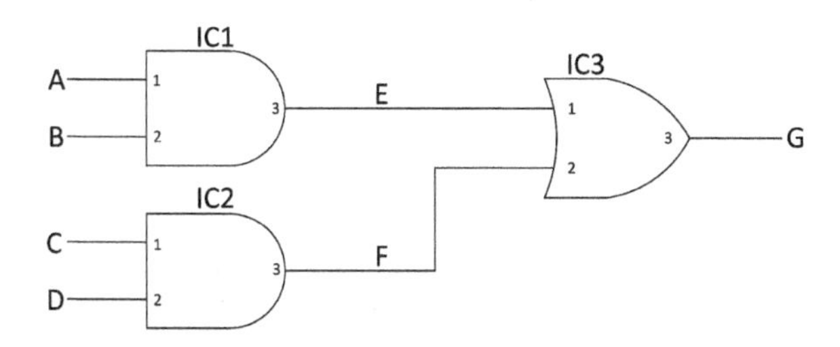

▲ **Figure 1.78** *Combinational logic*

Table 1.9 *Truth table for Figure 1.78*

A	B	C	D	E	F	G
0	0	0	0	**0**	**0**	**0**
0	0	0	1	**0**	**0**	**0**
0	0	1	0	**0**	**0**	**0**
0	0	1	1	**0**	**1**	**1**
0	1	0	0	**0**	**0**	**0**
0	1	0	1	**0**	**0**	**0**
0	1	1	0	**0**	**0**	**0**
0	1	1	1	**0**	**1**	**1**
1	0	0	0	**0**	**0**	**0**
1	0	0	1	**0**	**0**	**0**
1	0	1	0	**0**	**0**	**0**
1	0	1	1	**0**	**1**	**1**
1	1	0	0	**1**	**0**	**1**
1	1	0	1	**1**	**0**	**1**
1	1	1	0	**1**	**0**	**1**
1	1	1	1	**1**	**1**	**1**

Digital fault categories

As digital circuits use logic levels we are only concerned with incorrect logic levels. To make fault finding easier and more logical, there are only three types of digital faults that we normally consider for TTL circuits:

Logic output – Stuck at 0 (sa0),

Logic output – Stuck at 1 (sa1),

Logic input – Open circuit floating to a 1 (oc1).

Many faults could occur within a digital integrated circuit, caused by any one of the hundreds of transistors within the device failing. The net result of this failure is for the logic output to stay at logic 0 (sa0) or stay at logic 1 (sa1). Due to the sensitivity of digital circuits to damage by static electricity, logic inputs can sometimes be damaged and become open circuit (oc1). Note that in our circuit above, IC1.3 (sa1) would give the same fault symptoms as IC3.1 (oc1) and the effect on output G would be the same.

In Figure 1.78, the 4 inputs A, B, C and D can be (sa1), (sa1) or (oc1). IC1/2/3 output can be (sa1) or (sa1), and IC1/2/3 inputs can be (oc1).

Assume we are given the circuit above, and Table 1.10 with columns E, F and G_{GOOD} empty. We need to fill in the missing contents in the three empty columns. We can then see the difference between columns G_{GOOD} and $G_{BAD}1$, $G_{BAD}2$, $G_{BAD}3$ and $G_{BAD}4$. The columns with $G_{BAD}X$ contain faults, the possible faults are at at end of each column. If the $G_{BAD}1$ column given contained all logic '0' then the only possible fault within this circuit would be IC3.3 (sa0). If the $G_{BAD}2$ column given contained all logic '1' then possible faults within this circuit include: IC1.3 (sa1), IC2.3 (sa1), IC3.3 (sa1), IC3.1 (oc1) and IC3.2 (oc1). It would not be possible to determine which IC was faulty without making measurements with an oscilloscope, logic probe or a DMM. If the faulty data given was that contained in column $G_{BAD}3$ then possible faults would include Input A (sa1) or IC1.1 (oc1). If the faulty data given was from $G_{BAD}4$, then possible faults include IC2.3, Input C and Input D (sa0).

Table 1.10 *Truth table for Figure 1.78 with faults applied*

A	B	C	D	E	F	G_{GOOD}	$G_{BAD}1$	$G_{BAD}2$	$G_{BAD}3$	$G_{BAD}4$
0	0	0	0	**0**	**0**	**0**	**0**	**1**	**0**	**0**
0	0	0	1	**0**	**0**	**0**	**0**	**1**	**0**	**0**
0	0	1	0	**0**	**0**	**0**	**0**	**1**	**0**	**0**
0	0	1	1	**0**	**1**	**1**	**0**	**1**	**1**	**0**
0	1	0	0	**0**	**0**	**0**	**0**	**1**	**1**	**0**
0	1	0	1	**0**	**0**	**0**	**0**	**1**	**1**	**0**
0	1	1	0	**0**	**0**	**0**	**0**	**1**	**1**	**0**
0	1	1	1	**0**	**1**	**1**	**0**	**1**	**1**	**0**
1	0	0	0	**0**	**0**	**0**	**0**	**1**	**0**	**0**
1	0	0	1	**0**	**0**	**0**	**0**	**1**	**0**	**0**
1	0	1	0	**0**	**0**	**0**	**0**	**1**	**0**	**0**
1	0	1	1	**0**	**1**	**1**	**0**	**1**	**1**	**0**
1	1	0	0	**1**	**0**	**1**	**0**	**1**	**1**	**1**
1	1	0	1	**1**	**0**	**1**	**0**	**1**	**1**	**1**
1	1	1	0	**1**	**0**	**1**	**0**	**1**	**1**	**1**
1	1	1	1	**1**	**1**	**1**	**0**	**1**	**1**	**1**
						No fault	IC3.3 (sa0)	IC1.3, IC2.3, IC3.3 (sa1), IC3.1, IC3.2 (oc1)	Input A (sa1), IC1.1 (oc1)	IC2.3, Input C, Input D (sa0)

In each case we only consider one digital fault at any one time. Simultaneous multiple faults are very difficult to diagnose without computer assistance.

Interfacing

When connecting ICs to each other via short wires or tracks of a printed circuit board (PCB), as long as all of the ICs are of the same family, TTL, CMOS, ECL, etc., and the rules of fan-in and fan-out are followed, there are few problems that can arise. If you are connecting the output of an IC via a very long track, wire or backplane, then more consideration needs to be given to how we look after the integrity of the logic levels. If the signal path is longer, the logic levels degrade due to the capacitance and inductance of the track, wire or backplane. If the speed of the signal (clock rate or rise time/fall time) increases then the signal degrades. To optimise the signal and ensure that the logic levels received by the distant IC are valid and reliable we need to ensure that the driving IC has sufficient power output and that the signal path is of defined impedance. Digital circuit designers have to consider the logic of their design and the quality of the logic levels travelling through their system. Typical devices that are used to drive signals over long tracks, wires and backplanes are the 74ABTXXX series: ABT = Advanced BiCMOS technology; XXX identifies each unique device function. These devices can operate at high speeds, with considerable current output capability ensuring that logic signals arrive at their destination intact. If we were to connect a CMOS device to a TTL device, we would need to consider using an interface IC such as a CD4050B to connect the two devices together. To connect a TTL device to CMOS we find that a simple resistor connected to the junction of the two devices would ensure that the voltage levels expected by both devices are achieved. Both of these methods can be seen in Figure 1.79.

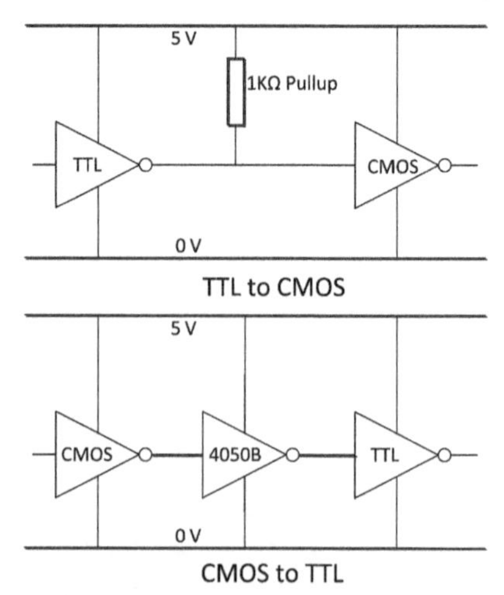

▲ **Figure 1.79** *Interfacing different device families*

Flip-flops (SR, JK, D)

We can connect AND, NAND, OR, NOR and NOT gates together to create more complex functions. We can connect an AND gate to a NOT gate to make a NAND gate. Similarly the OR and the NOT can be combined to create a NOR gate.

Connecting two NAND gates as shown in Figure 1.80 creates a cross-coupled latch. The truth table in Table 1.11 shows us that this device can 'remember' its output logic state (0 or 1).

It has problems when both inputs enter an illegal state (both inputs logic 0), the Q and not Q go to unknown output levels, that is, they could be either 1 or 0, no one knows. Careful circuit design is required to ensure these illegal states do not occur. It is more convenient to make the logic gate more complex to remove the possibility of these illegal states. Adding the additional logic can create a master–slave flip-flop which removes doubt when identical logic levels are applied to the inputs of the device.

One type of master–slave flip-flop is the JK. The JK solution to the problem of indeterminate outputs is to use a separate clock input. JK outputs do not change until the clock input returns to logic 0 or they are 'clocked out' on the falling edge of the clock, but the JK inputs are 'read' or 'clocked in' on the rising edge of the clock. A JK flip-flop symbol can be seen in Figure 1.81 with its truth table in Table 1.12. The point to note is that the Q and not Q outputs toggle (flip) when the clock input goes from low to high to low.

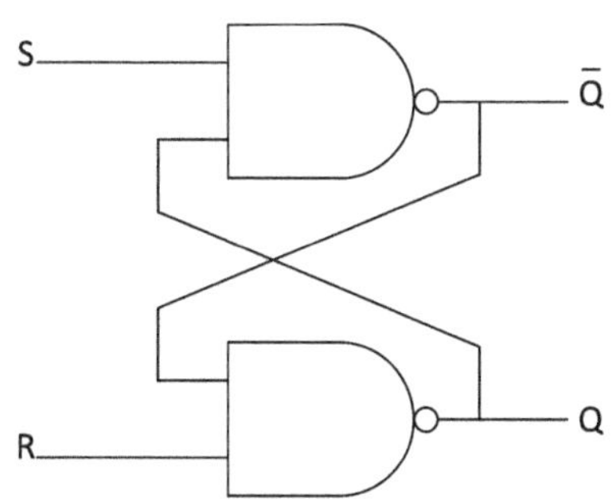

▲ **Figure 1.80** *Cross coupled NAND gates*

Table 1.11 *Truth table for cross-coupled NAND gates*

S	R	Q	/Q	Action
0	0	X	X	Outputs indeterminate
0	0	1	0	Set
1	0	0	1	Reset
1	1	No change	No change	Remember state

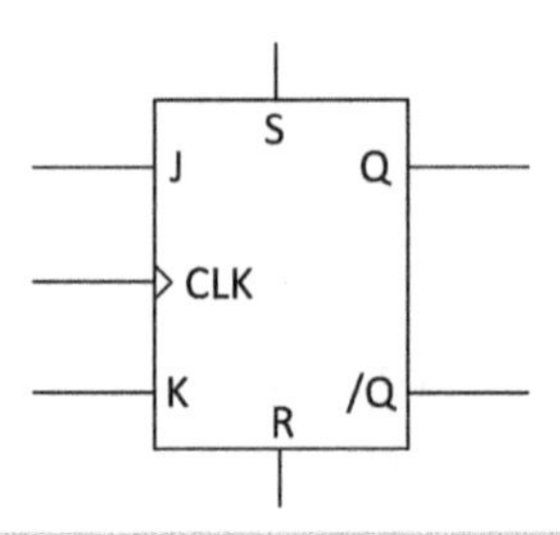

▲ **Figure 1.81** *JK flip-flop*

Table 1.12 *Truth table for JK flip-flop*

Inputs					Outputs AFTER clock	
J	**K**	**S**	**R**	**Clock**	**Q**	**/Q**
x	x	1	0	X	1	0
x	x	0	1	X	0	1
x	x	1	1	X	1	1
0	0	x	x	↑↓	No change	
0	1	x	x	↑↓	0	1
1	0	x	x	↑↓	1	0
1	1	x	x	↑↓	Outputs toggle	

x = do not care what the logic level is

A more popular logic device is the D-type latch. A D-type latch 'stores' the value of its D input usually after the rising edge of its clock. Figure 1.82 and Table 1.13 show the symbol and truth table. D-type latches and relatives are used because most logic design is 'synchronous'. That is, in an electronics circuit, every device that can 'store' logic levels is clocked simultaneously. By arranging for all devices to be clocked together, it eases the design burden and allows for timing issues to be addressed simply and helps to prevent 'races' from occurring. 'Races' are where a logic device receives two

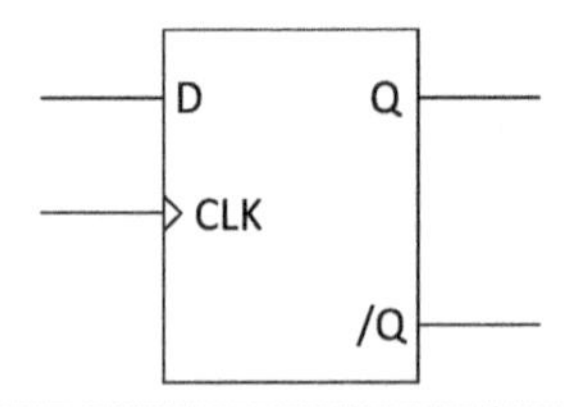

▲ **Figure 1.82** *D-Type latch*

Table 1.13 *Truth table for D-type latch*

Inputs		
D	**Clock**	**Q**
0	↑	0
1	↑	1
X	0	nc

nc = no change, x = don't care.

or more inputs from different sources at slightly different times. The logic device will demonstrate its confusion by sometimes latching in a '0' and sometimes a '1'. In digital design we need confidence that our circuits will always work in the expected way.

Synchronous and asynchronous counters

We can join memory elements (D-type latches or JK flip-flops) together to create more complex functions. A typical function that is used frequently is a counter. Connecting four D-type latches with some NAND gates creates a four-bit counter. The precise circuitry of the counter depends upon what facilities the counter is required to possess. Four-bit counters can count from 0 to 15 (16 clock pulses) or 0 to 9 (reset after 10 clock pulses). Some counters can be made to load a number between 0 and 15 and count from there. Figure 1.83 shows typical content of a 74LS16X, a synchronous four-bit counter with parallel load pins. The X identifies the unique counter: 0, 1, 2 or 3.

Analogue to digital converters

The world around us is analogue in nature: the size of a signal, the speed of a machine, the temperature are all varying values. A computer or other logic device works predominately with digital or binary signals. Analogue to digital converters (ADCs) convert a varying voltage into a digital, binary value. There are many different methods used to convert an analogue voltage to a digital representation within ADCs.

In Figure 1.84 we can see a simple ADC block diagram. As the voltage to be sampled could be constantly changing we need to take a snapshot of the voltage and freeze the value during the conversion process. The voltage to be sampled is fed via an amplifier to a switch. The switch is controlled by a pulse generator. The pulse generator controls the sequencing of the complete device. A pulse from the pulse generator closes the

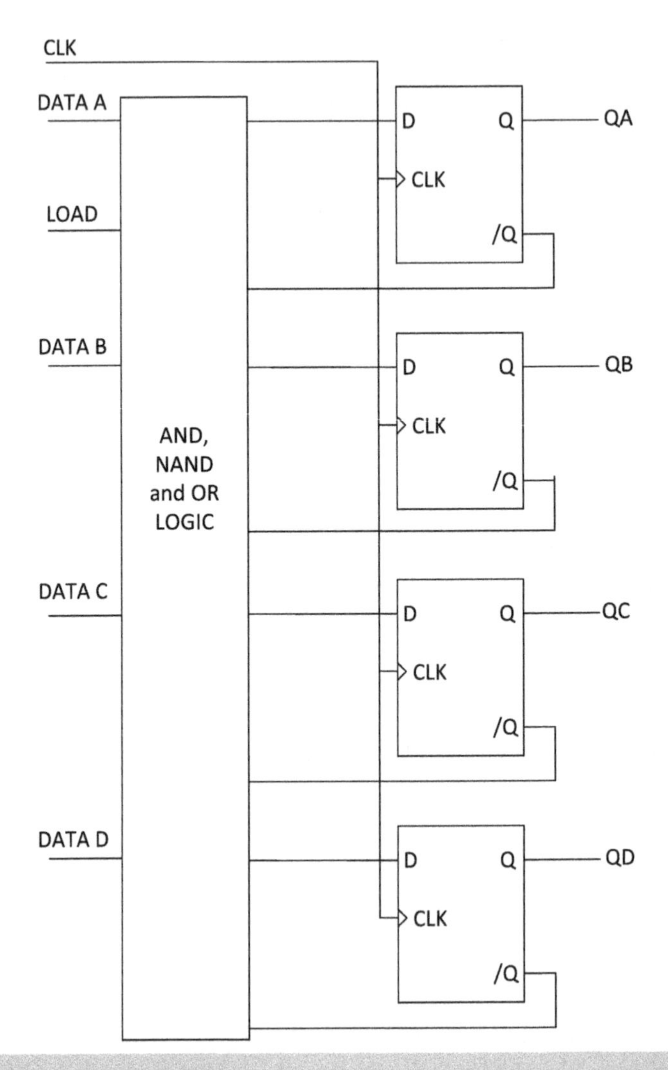

▲ **Figure 1.83** *Synchronous 4 bit counter*

switch and for a fraction of a second the capacitor charges up to the same voltage as the input voltage. The pulse generator removes the pulse to the switch, isolating the capacitor. The capacitor now contains the same voltage that was applied to the input pin of the ADC. The pulse generator now sends a stream of high speed clock pulses to the integrator and the binary counter. Each pulse received by the integrator causes the integrator voltage to be increased by a fixed amount. Simultaneous to this, a pulse to the counter increments the counter by one. The voltage from the capacitor and the integrator are both fed into the comparator circuit. The comparator stops the pulse generator when the voltage from the integrator is greater than the voltage from the

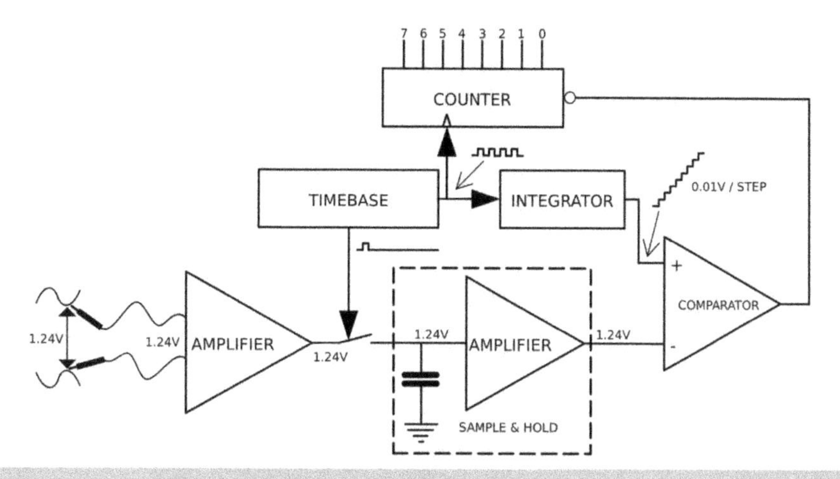

▲ **Figure 1.84** *Analogue to Digital Converter (ADC)*

capacitor. When this occurs the counter contains a binary number that represents the analogue voltage on the input pin.

Digital to analogue converters

A digital computer can process binary information at a very fast rate, but many devices in the real world are analogue and necessitate a digital to analogue converter (DAC) to enable a computer to control or communicate with analogue devices. In Figure 1.85 we see a simple block diagram of a DAC. The computer sends a binary word to the DAC, the DAC captures this word. Each bit of the word is allocated a weight or value.

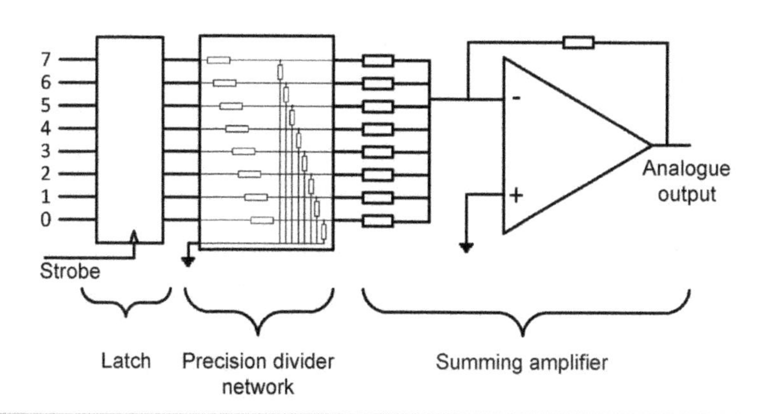

▲ **Figure 1.85** *Digital to Analogue Converter (DAC)*

Table 1.14 *Binary input to DAC*

				MSB				LSB	Value
Bit Name	7	6	5	4	3	2	1	0	Value
Weight	128	64	32	16	8	4	2	1	
	0	0	0	0	0	0	0	0	0
	0	0	0	0	0	0	0	1	1
	0	0	0	0	0	0	1	0	2
	0	0	0	0	0	0	1	1	3
	1	1	1	1	1	1	1	1	127

In Table 1.14 a list of bit positions and their respective values is shown. If, in the binary word, a bit is set or equal to a logic one, the value of that bit is added to the running total output voltage. This summing operation occurs the moment the binary word from the computer is captured by the DAC. The output voltage of the DAC is the weighted sum of all of the binary bits.

RAM and ROM memory devices

Digital computers require memory to operate. They require different types of memory to suit a range of tasks. As the computer requires a computer program to be available the instant power is applied to the computer system, we need a memory device that retains its data even when the computer system has lost power. This type of memory device is called a Read Only Memory device (ROM). The data within the device remains even when the ROM device is without power. When the ROM device is provided with power, the program data within it is immediately available. The ROM device cannot be changed easily and is ideally suited to hold the first program that a computer runs at power up. (The ROM device in a personal computer (PC) contains a program that is called Basic Input Output System (BIOS) and is the first program that your PC runs when you switch it on.)

Random Access Memory (RAM) serves a different function. It allows the computer to store answers to calculations and variable information while a computer program is running. RAM devices allow both reading and writing. The downside, they forget all the data when the power is removed to the device. In modern computers large amounts of RAM are provided to allow computer programs to be loaded, which results in faster program operation.

Memory technology is changing rapidly. ROMS have made way to become Programmable Read Only Memory (PROM), which in turn allowed for the development of Erasable Programmable Read Only Memory (EPROM). EPROMs spurred the development of EEPROMS (electrically EPROMS) sometimes called EAPROMS (electrically alterable PROMS) or E^2ROMs. FLASH EPROMS allow easier data updates. RAMS manufactured using CMOS technology can, with a small back up battery, retain all data. This is how laptop computers can hibernate.

Displays

Liquid crystal display

Liquid crystal display (LCD) displays allow flat panel display screens to be constructed. Liquid crystals are microscopically small particles suspended in a liquid. The crystals are manufactured to point in one direction. If a small voltage is applied to the liquid crystal the crystals point in a different direction. By sandwiching the liquid crystal fluid between two polarized filters at 90° to each other, light can pass through one filter, through the liquid due to the alignment of the crystals, then through the last filter. When a voltage is applied across the liquid crystal the crystals change direction and block the light path. By making small the area where the liquid crystal is used and shaped to create numbers, LCD displays can be used to present numbers to users. LCD displays can reflect light from a reflective surface after the last filter. This type of LCD requires ambient light to operate. Other LCD displays contain a 'backlight' located after the last filter where the reflector would be. Use of a 'backlight' enables an LCD to be used in low light areas (Figure 1.86).

By making the area of liquid crystal very small, fitting many thousands into a small area, graphic style displays can be constructed. Each liquid crystal area or pixel can be individually controlled. Higher voltages cause brighter or darker images to appear on the front of the LCD screen. LCD screens do not work at low temperatures as the crystals cannot move when frozen or below critical temperatures.

LCD monitor screens have 'cold cathode' backlights that spread the light level evenly across the display screen. The backlight is always on when the monitor is in use, using considerable power. LED backlighting allows for the screen to be illuminated by thousands of LEDs. LED screens still use LCDs as the main on/off light control mechanism to allow the image to be seen on the screen. OLED, or organic LEDs, allow each LED cluster to be turned off individually, giving much better 'black levels'.

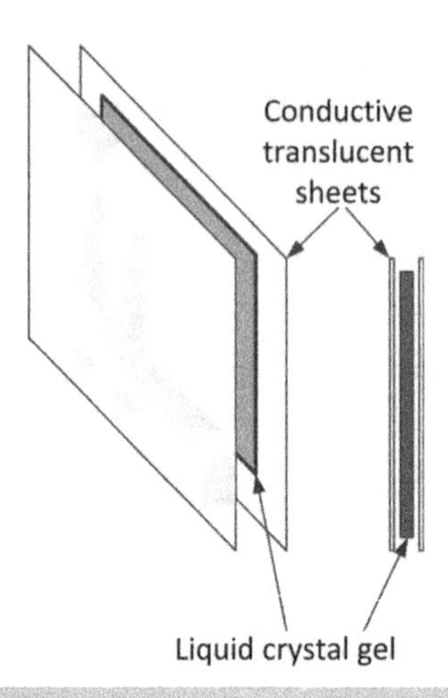

▲ **Figure 1.86** *LCD panel construction*

Cathode ray tube

A cathode ray tube (CRT) works as follows:

On the right of Figure 1.87, a heating surface called the cathode is heated until white hot, emitting a cloud of electrons. On the left of Figure 1.87 is a glass plate, coated in a fluorescent material that glows when hit by electrons. A high positive voltage (2,000 V) is applied to the fluorescent coating causing the cloud of electrons around the cathode

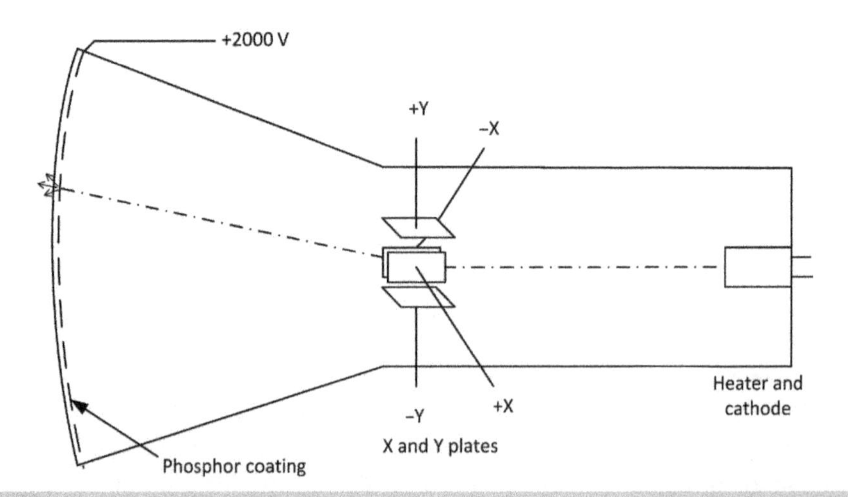

▲ **Figure 1.87** *Cathode ray tube*

to be pulled at high speed from the cathode to the fluorescent material. When the electrons hit the fluorescent screen, light is emitted. The point of light on the screen can be made to move, either left or right and up or down, by the application of small voltages fed to the X and Y plates. A positive voltage on one X plate and a negative voltage on the opposite X plate will cause the beam of electrons to be pulled towards the positive X plate and repelled from the negative X plate. Any bending of the electron beam will cause the spot of light on the screen to move accordingly. To make the CRT a useful device to display electrical signals, it is usual to arrange for the electron beam and therefore the point of light to be swept from left to right along the X axis. This would cause a line to appear on the screen. Any signals then applied to the Y plates would cause the spot of light to move up or down simultaneously with the X axis movement. To get the point of light to move along the X axis in a regular manner, a saw tooth waveform is applied to the X plates. Use of the CRT is described in the section on Oscilloscopes.

Measuring Instruments

Universal meter

The most useful electrical measuring instrument for maintenance and fault finding is the digital multimeter (DMM). The DMM allows us to conveniently measure voltage, current and resistance. Some models allow us to measure additional parameters such as temperature, capacitance, etc.

The modern DMM is based on an ADC changing the analogue voltage being measured into a digital value for display. In Figure 1.88 we see a simple arrangement of inputs and range selection on the left, converters in the centre feeding function switches (choice of AC, DC, ohms or current) feeding onto the A/D converter and then the display device. The range selection can be manual or automatic and changes high voltages to lower values to keep within the range of the A/D converter. The ohms converter converts the current flowing through the test subject into a voltage. The current shunt converts the measured current into a small voltage for the A/D. The function switches are usually manual and may be combined with the range switches.

It is important to note that just because the DMM appears to generate extremely accurate reading, for example, 2.49999 V, the true measured value could be between a range of values, depending upon the stated accuracy of the instrument. Figure 1.89 shows a modern, handheld DMM by Fluke.

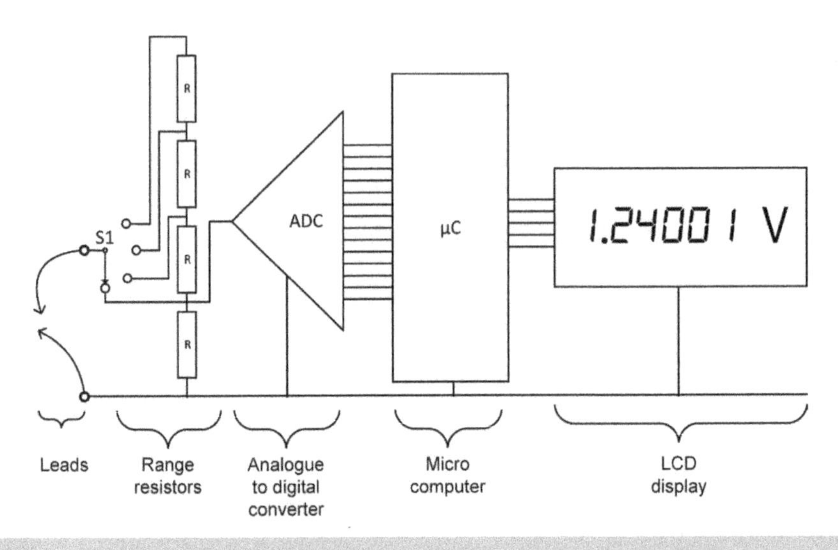

▲ **Figure 1.88** *Digital multimeter*

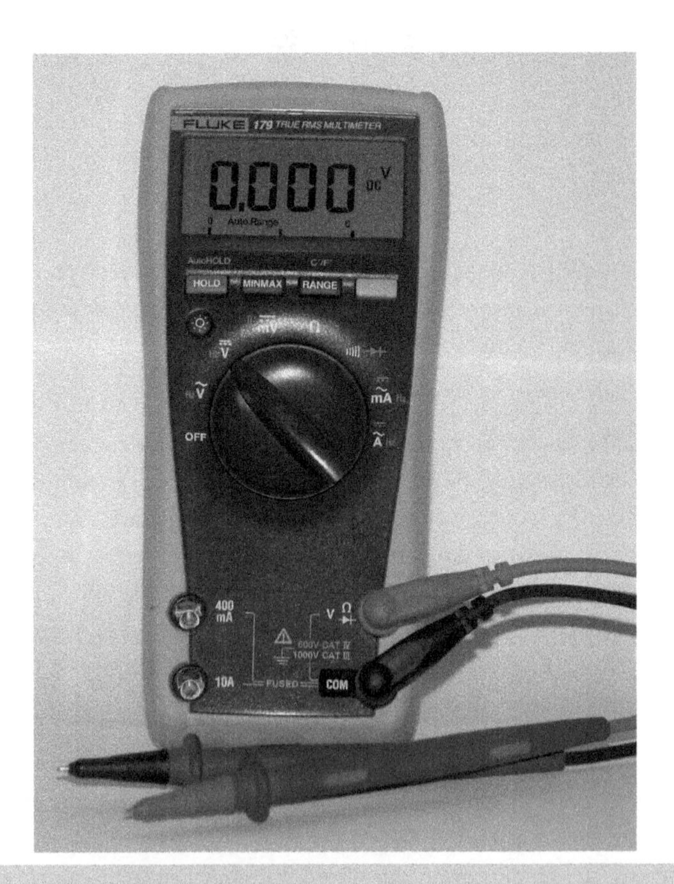

▲ **Figure 1.89** *Fluke 179 Digital Multimeter*

Oscilloscope

Oscilloscopes allow you to see and measure electrical signals. This is especially useful when electrical signals are changing rapidly. A DMM will not follow and measure rapidly changing signals. An oscilloscope has a method of connecting to the signal to be viewed, and a method of displaying the signal to the user. The display screen is available in two major technologies:

1. Cathode Ray Tube (CRT)
2. Liquid Crystal Display (LCD)

The CRT has been described earlier. An oscilloscope can utilise a CRT to display electrical signals in the following way. Figure 1.90(a) shows that with 0 V applied to all inputs of the CRT, a bright stationary spot will be displayed in the centre of the screen. The signal to be displayed is connected to the vertical or Y plates. When a signal voltage is non-zero, the bright spot on the front of the CRT screen will move away from the centre line, for example, apply 1 V to the vertical plates and the spot moved 1 cm upwards from the centre line, apply –1 V and the spot moves to 1 cm below the centre line (see Figure 1.90(b) and 1.90(c)). In this way, the applied static signal can be measured. If we were to apply a sine wave signal to the Y plates we would see a vertical line as seen in Figure 1.90(d). We would not be able to determine that the signal was a sine wave. To see moving or changing signals, we apply an increasing voltage to the horizontal or X plates. This has the effect of moving the bright spot from left to right across the screen, as seen in Figure 1.90(d). It can be seen that if you were to apply an increasing voltage onto the X plates simultaneous to the application of a sine wave voltage to the Y plates, we would see the bright spot on the front of the CRT trace out a sine wave, shown in Figure 1.90(f). Oscilloscopes need to be able to amplify small signals to make them visible on the screen, and to be able to attenuate large signals to make them fit onto the screen. The vertical amplifier is used to perform this task. The vertical amplifier is a variable gain amplifier and has high input impedance.

To connect the oscilloscope to the point of measurement (transistor, resistor, capacitor, etc.) we need a cable that allows the oscilloscope vertical amplifier to receive signals. The cable and connection device is called an oscilloscope probe or scope probe for short.

Each scope probe is slightly different in length and capacitance. The vertical input socket capacitance value of each oscilloscope is slightly different from the other. Due

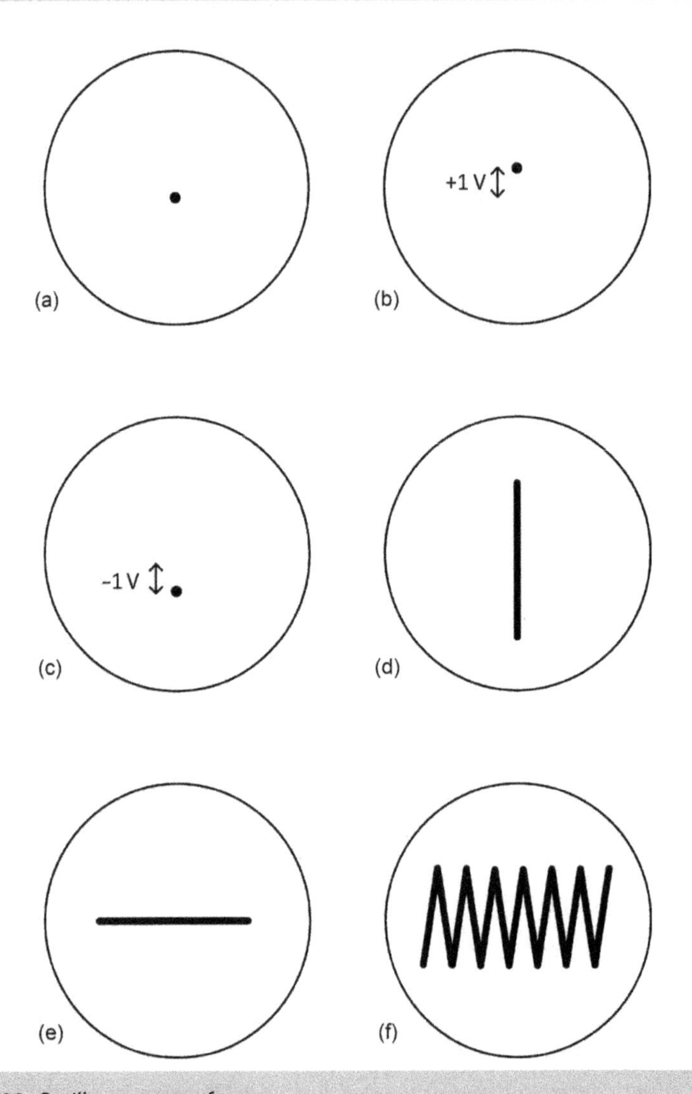

▲ **Figure 1.90** *Oscilloscope waveforms*

to the high input impedance of the vertical amplifier and the fact that we can display very small voltages, and the difference in capacitance, it is important that the scope probe capacitance is correctly adjusted (compensated) to match that of the vertical amplifier. To adjust a scope probe so that it matches correctly the input impedance of the vertical amplifier, we make use of the small capacitor fitted inside the body of the scope probe. To perform the adjustment, we connect the scope probe and earth to the calibrated test output on the front panel of the oscilloscope. We adjust the oscilloscope controls until we can see a reasonable sized square wave. We then adjust the compensation capacitor in the probe body until we get close to a perfect square wave. The oscilloscope and probe combination are then ready to be used (Figure 1.91).

Scope probes can be just a short section of wire or a complex device. It is customary to use a times ten scope probe (×10 probe). A ×10 probe has the ability to multiply the input impedance of the vertical amplifier by a factor of 10, but in doing so, reduces the voltage into the vertical amplifier by a factor of 10. Whenever we use a ×10 probe to make a measurement, we have to remember to multiply any voltage reading we see on the oscilloscope screen by 10. The vertical axis of the oscilloscope and the gain adjustment are marked in volts per centimetre. Oscilloscopes need to be able to display signals of different frequencies, low and high. To enable the oscilloscope to display low frequencies, the signal we create and send to the X plates needs to be a low frequency. If we wish to display a higher frequency, we need to increase the frequency of the signal we place on to the X plates. This gives rise to the need to be able to adjust the frequency of the signal that we inject onto the X plates. The horizontal axis of the oscilloscope and the horizontal speed/frequency adjustment is marked in seconds per centimetre.

To enable the oscilloscope to conveniently display dynamic signals (sine waves, square waves, etc.) requires that we freeze the horizontal axis and release it only when the signal has reached a certain voltage. By varying the level of this trigger voltage we can arrange for different parts of the waveform to be displayed. Adjustment of the trigger voltage requires much practice, as some waveforms that we wish to display are not

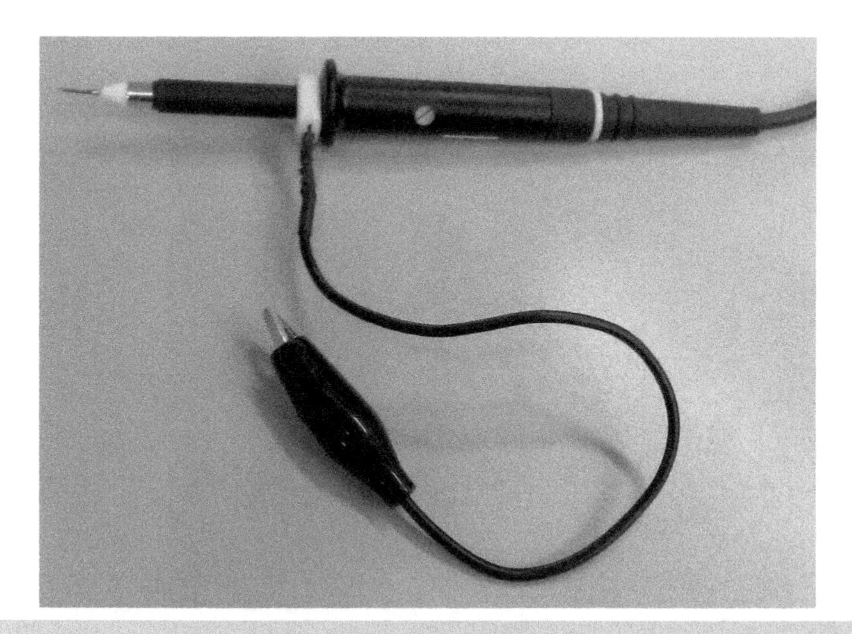

▲ **Figure 1.91** *Oscilloscope probe*

very stable, particularly when a fault exists in the circuit under test. Figure 1.92 shows the block diagram of a simple oscilloscope.

Figure 1.93 shows the front panel of a modern digital oscilloscope. Digital oscilloscopes digitise the measured signal to allow complex measurements to be made upon the signal. They can use a modern display technology like LCD, removing the need to use a heavy CRT.

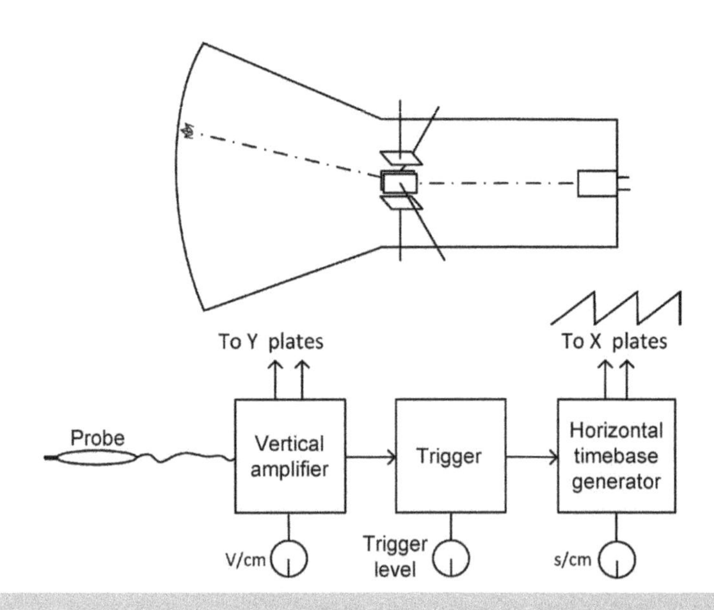

▲ **Figure 1.92** *Oscilloscope block diagram*

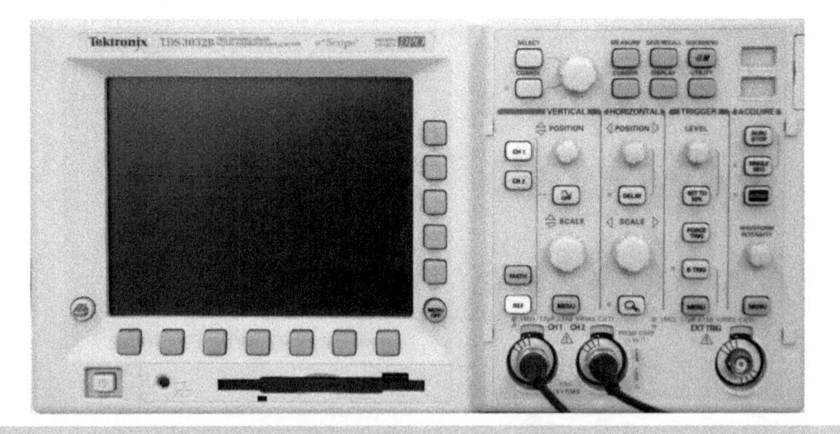

▲ **Figure 1.93** *Tektronix TDS 3032B Oscilloscope*

Logic probe and pulser

When an oscilloscope is not available to investigate AC conditions within a circuit, a Logic Probe can be utilised. A Logic Probe is a small handheld device that allows digital voltages and pulses to be detected (Figure 1.94).

A Logic Probe consists of electronic circuitry housed within a compact hand held case. The housing contains a sharp probe which is used to connect to the circuit under test. The Logic Probe is powered by the circuit under test. The housing contains LED indicators which show the voltage detected at the probe tip. LEDs illuminate for logic 0, logic 1 and pulses.

The Logic Probe can be used on CMOS and TTL voltage circuits. A sophisticated probe can detect appropriate voltage thresholds; other Logic Probes have a switch to select a digital family.

Logic Probes are used to quickly ascertain activity within a digital circuitry. By connecting the probe tip to an output of a logic gate, it can be determined if the logic gate is driving a logic 0 or logic 1. The destination of the logic signal can be checked. By probing logic outputs and inputs, logic levels and pulses can be followed around a circuit, helping you to reach a diagnostic conclusion.

When a digital circuit is completely static with no component pins changing logic level, it can sometimes be difficult to find some faults. We can use Logic Pulsers to assist (Figure 1.95).

▲ **Figure 1.94** *Logic Probe*

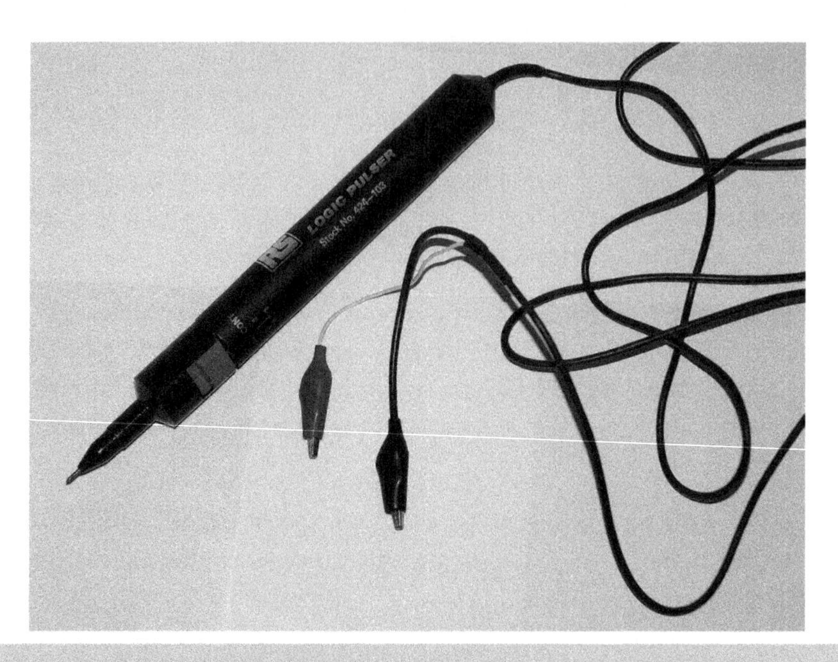

▲ **Figure 1.95** *Logic Pulser*

Logic Pulsers are small, handheld electronic devices that can deliver a short logic pulse to a digital circuit. When using a Logic Probe to check the logic levels within a digital circuit it can sometimes be beneficial to cause some activity which can be monitored with a Logic Probe. Logic Pulsers are powered by the circuit under test and contain a button that when pressed injects one pulse into the circuit. The Logic Probe injects the pulse taking into account the pre-existing logic level. So a circuit that had logic 1 would be injected with a low going pulse. That injected pulse can then be traced through the PCB track or to a digital device with the aid of the Logic Probe. Logic Pulsers can contain a button that causes a stream of pulses to be injected. The length of a Logic pulse is deliberately kept short: typically <20 ns with a long *off* period to avoid damaging digital outputs. A long *on* pulse would cause any connected digital outputs to overheat excessively.

Circuits

The ability to navigate around an electronics circuit is a valuable skill. We demonstrate how to determine voltages within a circuit, which will enable us to find faulty components. Figure 1.96 shows a number of components connected in a configuration.

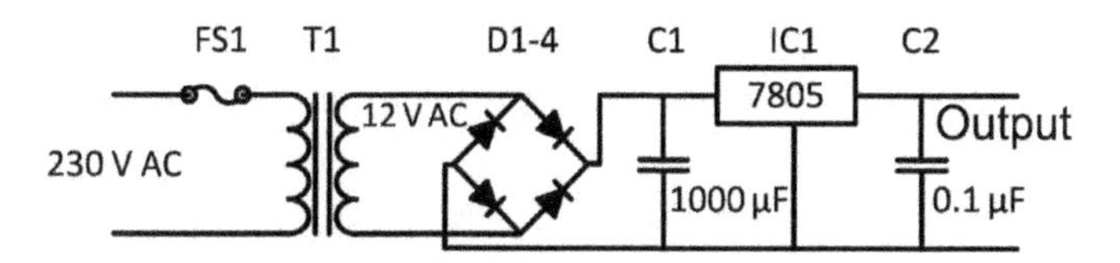

▲ **Figure 1.96** *DC Power Supply Unit (PSU)*

To understand what function this circuit performs we need to understand what each component does. Working from left to right of Figure 1.96, FS1 is a fuse used to protect the wiring to this whole circuit; it offers no protection to this circuit. T1 is a mains step-down transformer; we can tell this because 230 V AC is applied to the left hand side, and 12 V AC is shown on the right hand side of the transformer. D1 to D4 is a standard configuration for a bridge rectifier. This provides full wave rectification of the 12 V AC into DC. This DC voltage has a very large ripple voltage so capacitor C1 is provided to smooth the pulsating DC into a relatively smooth DC voltage. IC1 is an industry standard three pin voltage regulator; 7805 is a 5 V, 1 A regulator with short circuit protection. The output voltage can usually be determined from the component name. C2 is required to improve the output characteristics of the 7805 and all manufacturers' datasheets recommend the use of a 0.1–1 µF capacitor in this position. The above points to this circuit being a 5 V DC power supply powered from a 230 V AC supply with a maximum 1 A output current.

In Figure 1.97 we see a transistor drawn 'on its side' connecting a battery to a light bulb. On the left we have a 24 V battery and on the right, a 9 V lamp. Transistors have approximately 0.6 V between their base and emitter pins when they are conducting. We know from Figure 1.27 that Zener diodes have approximately a constant voltage across them when they are used in their Zener region. Putting this collection of parts together we get a very simple power supply that allows a 9 V lamp to be connected

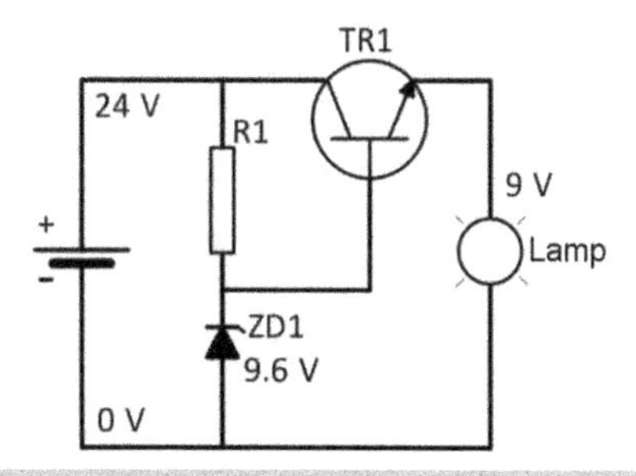

▲ **Figure 1.97** *Simple stabilised supply*

to a 24 V battery. The Zener is chosen to have a 9.6 V operating point. The transistor, when conducting, has the 9.6 V from the Zener diode, causing the transistor emitter to follow at 9.0 V (0.6 V less than the base voltage). If the lamp was changed for a device consuming more current, the transistor would allow more current to flow though itself, maintaining its emitter voltage at approximately 9.0 V.

The transistor is operating in 'series pass' mode, all the current consumed by the load (lamp) has to pass through the transistor, causing the transistor to dissipate a large amount of power in the form of heat.

For example, if the lamp was a 10 W device, the current flowing through it would be:

$$P = VI \quad \therefore I = \frac{P}{V} = \frac{10}{9} = \underline{1.11\text{A}}$$

The power dissipated by the transistor would be:

$$P = VI = (24\text{V} - 9\text{V}) \times 1.11\text{A} = \underline{16.65\text{W}}$$

So we waste approximately 17 W to energise a 10 W lamp, a total of 27 W. Not very efficient!

2

NAVIGATIONAL AIDS – THEORY AND FAULT FINDING

Micro Computers

Most, if not all, modern navigational equipment on board merchant vessels is computer controlled. An understanding of microprocessors will be helpful when fault finding modern equipment.

Many types of microprocessors are available and equipment makers can choose from a range of hundreds; 8, 16, 32 and 64 bit processors give equipment manufacturers a range of processing power to suit different requirements.

The amount of processing power required to control a typical electronic system on board a ship can be provided by an 8 or 16 bit processor. A typical 8 bit microprocessor consists of a number of functions which include:

- ALU – Arithmetic and logic unit
- ACC – Accumulator
- Registers

- PC – Program counter
- Instruction decoder

The microprocessor will be fed with a clock signal, typically a square wave; the frequency could be between 4 and 100 MHz. When the microprocessor is powered up and receives its first clock signal, its internal design is such that it tries to obtain its first user instruction. This instruction (part of the equipment computer program) must reside somewhere the microprocessor can access it. A Read Only Memory (ROM) device provides a place where the microprocessor can find instructions. A ROM device holds the list of instructions that we want the microprocessor to obey. A ROM device holds its contents even when the power is removed. When the power is reapplied, the program instructions will be available to the microprocessor. As the microprocessor receives clocks, it reads an instruction from the ROM, decodes the instruction, and then obeys the instruction. Each instruction is a binary pattern of ones and zeros. It is not necessary for us to understand these instructions to find faults with computer controlled equipment. We need to have an understanding of the system architecture to enable us to understand how certain faults can give rise to certain failures.

For a microprocessor to function it needs a series of clock pulses; square wave in nature, the voltage limits of the clock signal match the technology of the microprocessor. Older devices require TTL 5 V clocks; more modern devices require 3.3 V or lower. The microprocessor will, when powered and fed with a clock, read and obey instructions. That is all it can do. The power of computers lies in the sequence of binary instructions. The microprocessor reads the first instruction, decodes it, obeys it, increments the program counter, then gets the next instruction. This loop continues until the power is removed or a fault occurs.

In Figure 2.1, we see a simple microprocessor device containing a minimum of functions. The clock controls the timing and sequencing of the whole device. It has eight data lines with which it can read and write from and to memory. Sixteen address lines enable selection of appropriate addresses. The Program Counter (PC) is connected to the address bus, so whatever their content, the PC always contains the next address to be accessed. The data bus can be read and the data value can be moved into the Accumulator (ACC). Data can also be written from the ACC to the data bus. Some instructions force the Arithmetic and Logic Unit (ALU) to manipulate data in the ACC and place the result back into the ACC. For a simple microprocessor to work correctly it requires a number of support devices: ROM, RAM, IO and clock devices as a minimum. So frequent is the need for these support devices that device manufacturers now include this functionality within the main device. When this additional functionality is added, the name of the device becomes microcomputer.

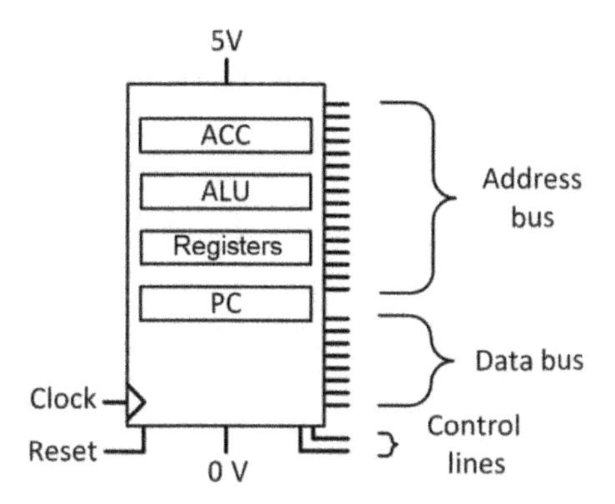

▲ **Figure 2.1** *Microprocessor*

A microcomputer contains (in many cases) all electrical parts required. Only power need be added, to create a fully functioning computer system.

System manufacturers decide whether to use a microprocessor or microcomputer within each design; it is a design choice.

An important device included in many microcomputers is a Watchdog timer device (WDT). A WDT enables a computer system to recover from temporary faults. When a microcomputer experiences a set of inputs that no one predicted during the design of the system, it could get stuck in an unexpected loop and appear to freeze. To cover this eventuality a WDT is used. The WDT is periodically pulsed (or kicked) by the computer program. If the WDT is repeatedly kicked, it does nothing, but should the computer program get stuck in an unexpected loop, this kicking process will cease, and the WDT will, after a few milliseconds, reset the microcomputer. All professional computer controlled equipment will have a WDT. If you monitor the reset input pin on a modern microprocessor/micro controller device and observe it pulsing, it can be due to a fault within: the processor, the memory (RAM or ROM), an input/output (IO) device, data bus or address bus. Any loss of functionality can cause a WDT to time out and reset the processor.

Before we look at compete computer based systems let us look at how the computer communicates with the outside world. IO devices read and write binary bits of information and send or receive these 'bits' from the 'outside world'. These binary bits of information can be presented to the outside world in two ways, parallel or serial. In parallel mode, 8 bits can be written from the microprocessor or computer to an 8 bit parallel IO device, these 8 bits would then be available to the outside world to control equipment or energise a relay. The converse can happen in the opposite direction which would enable the computer program to 'sense' if a switch or button in the 'outside world' had been operated by reading the binary level from the switch into the ACC. Parallel inputs and outputs work well up to medium data speeds. High speed data transfer becomes difficult in parallel due to problems with ensuring each of the 8 or 16 signal wires change state at exactly the same time. Differences due to manufacturing tolerances can cause one or two of the bits to be slower than the others. Due to this reason, serial more than parallel communication is used.

Original Equipment Manufacturers (OEM) tend not to use desktop computers to control their systems. Whether a RADAR or AIS, the OEM will choose a single board computer (SBC) to control the system. The SBC contains the processor, memory and plentiful Input/Output (IO) to interface and control the rest of the system. This is good from the servicing point of view. The SBC will be dedicated to running one system. Its program will be fixed in firmware. It will run self tests at switch on and these tests can usually be invoked at will from the control panel to help us diagnose faults. SBCs come in all shapes and sizes dependent upon how much computing power your system requires and how much IO.

Serial communications

Serial is slightly more complicated. If we only have one wire instead of 8, and we still had 8 bits of binary data to transfer, then serialising the 8 bit data, sending along the wire one bit at a time, would be serial transmission.

Since the serial receiver does not know when the computer is going to start transmitting the serial data, we need to add a 'start bit' to the 8 bits of binary data. If the receiver 'knew' how quickly we were going to send each bit of data, the receiver could be waiting to check each bit to see if it is a one or a zero. After all of the bits have been transmitted, the receiver will now have an exact copy of the transmitted binary data. To help ensure that no data is corrupted in transit, a parity bit can be added to the 8 bits of binary data. Prior to sending, the 8 bit value is checked to see whether the number of 'one' bits is even or odd in value. Depending on the settings, we add another 'one' to make the count of 'one' bits to be either odd or even. The receiving end, on receipt of this parity bit, rechecks the count of 'one' bits, compares with the agreed rules and raises an error if there is a mismatch.

All of this serialising and de-serialising is processor time intensive and a very common task. System designers tend to use dedicated serial IO devices that perform this task automatically. This leaves the system designer to spend more time on the remainder of the system.

Computer controlled systems make use of Universal Asynchronous Receiver Transmitter (UART) devices such as the 6402, 8251, 16550 and derivatives, which perform the conversion from binary to serial and back without the computer or the designer concerned with the detail. Both the transmitting system and receiving system have

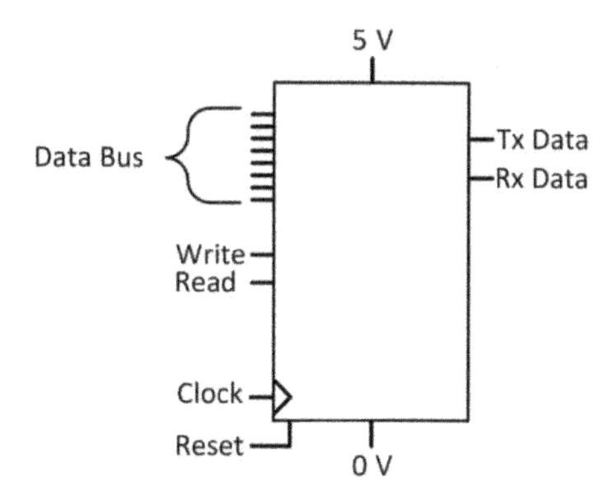

▲ **Figure 2.2** *Universal asyncronous receiver transmitter (UART)*

to use the same data rate. Typical values for communication between small computer controlled systems include 4800, 9600, 19200 baud. Baud refers to bits per second including the start, stop and parity bits (Figure 2.2).

If the serial information is to be transmitted more than a short distance or is to be transmitted to another system, it is wise to ensure that the serial link is robust and reliable. To achieve reliable and robust transmission of serial data, a number of techniques have been developed.

Figure 2.3 shows three serial communication links that could be used to send data from one item of equipment to another, such as GPS to AIS. The first is RS232, which uses one signal wire with one return wire. The voltage levels used by a RS232 transmitter circuit are typically ±12 V. Due to the use of only one signal wire, RS232 links are prone to noise pickup and have a limited distance of operation.

RS422 shown in Figure 2.3 overcomes much of the noise problem of RS232 by two signal wires transmitting equal but opposite polarity voltages (±6 V). Use of a balanced transmission system increases the maximum range of operation and greatly reduces noise pickup.

RS485 uses the same voltage levels of RS422 but adds the ability of disabling transmitters. This allows multiple transmitters the ability to drive signals along a cable to multiple receivers; each transmitter being enabled just prior to its use in transmission.

The common point of the above links is the copper connecting wire between the transmitting system and receiving system. In some cases it may be advantageous to have total electrical isolation (galvanic isolation) between equipment within a system. We achieve this by using the NMEA 0183 system.

Figure 2.4 shows a typical NMEA 0183 isolated transmission system. The interface makes use of opto-isolators which consist of a LED and a photo sensitive transistor

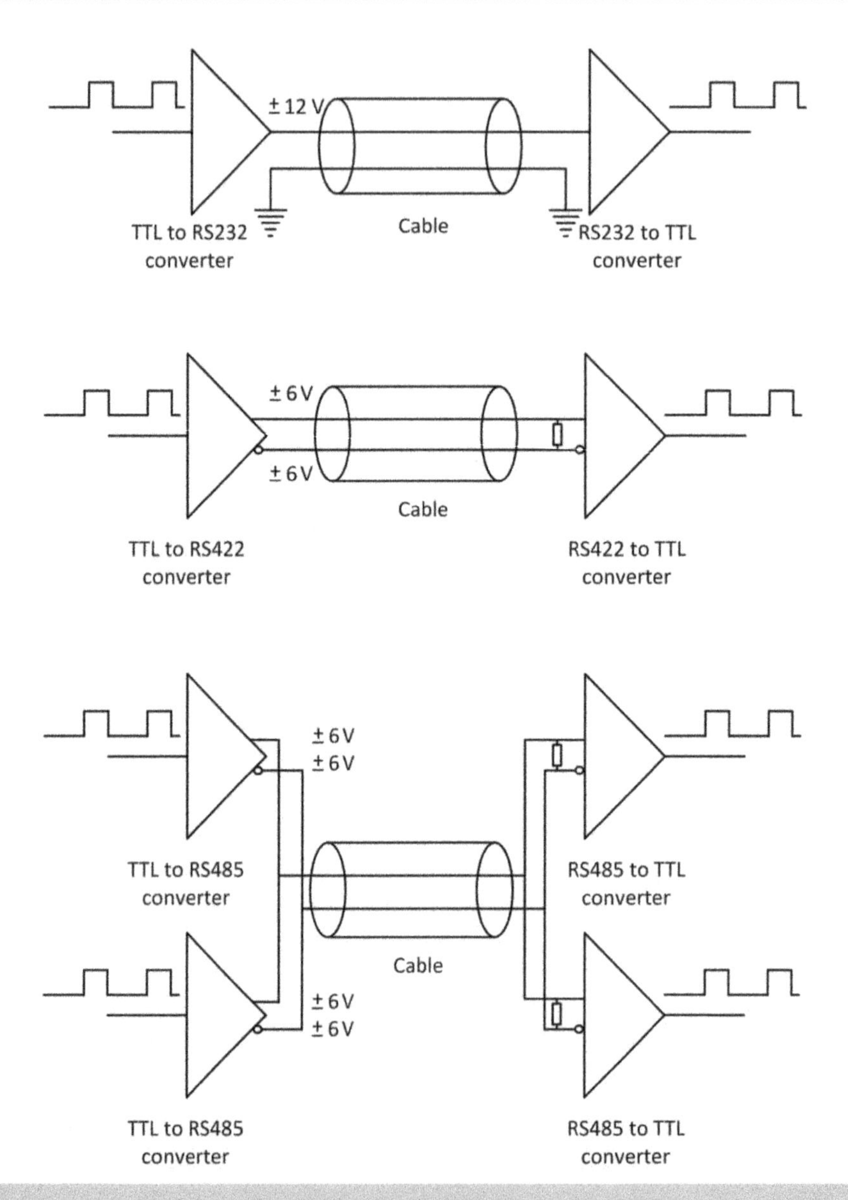

▲ **Figure 2.3** *Simple serial links – RS232, RS422 and RS485*

within a single device. There is an air gap between the LED and transistor of sufficient distance to prevent a high voltage flashover. A typical device, a SFH6186, is designed to withstand 5300 V without failing. By using opto-isolators we can ensure that any failure within one connected equipment will not be passed along the inter-system cabling and damage a connected equipment. It should be noted that the use of two opto-isolators on the transmission side causes the voltage levels to be opposite in polarity to each other, this gives NMEA 0183 the advantages of RS422/RS485 for long distance operation and good noise performance.

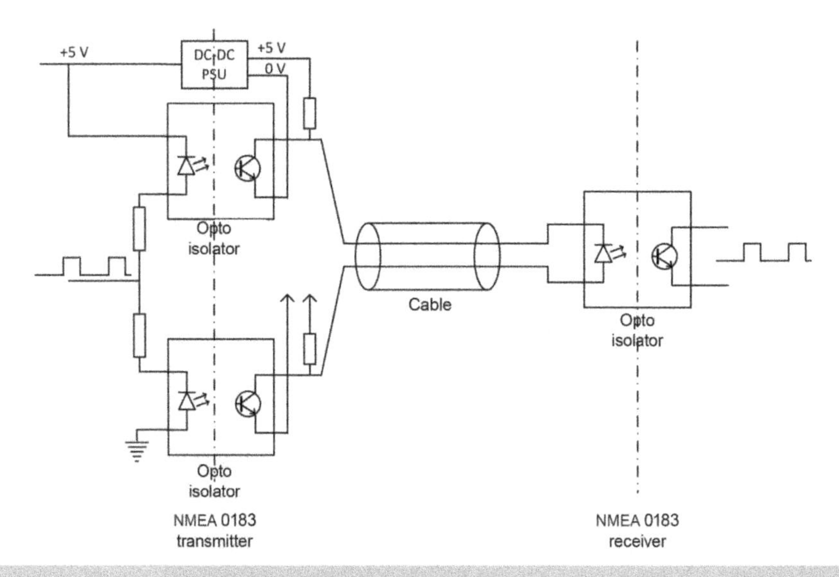

▲ **Figure 2.4** *NMEA 0183 isolated communication link*

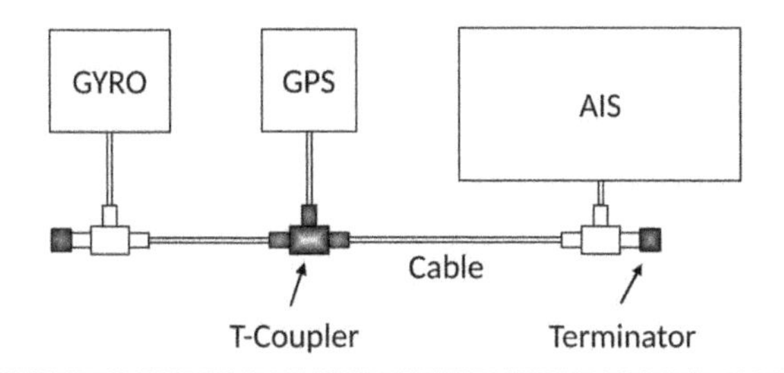

▲ **Figure 2.5** *NMEA 2000 system*

With vessel complexity increasing, and the desire to install more navigational electronics, NMEA introduced the NMEA 2000 system. This aims to bring the benefits of 'plug and play' techniques in the computer world into the marine electronics world.

Figure 2.5 shows a typical small system of navigational aids connected via the NMEA 2000 system. NMEA 0183 did not control how the physical wiring should be undertaken, but NMEA 2000 specifies all parts of the network, even down to type and size of connectors. Also, NMEA 2000 is a networking system where up to 50 devices can be connected together on one cable. The underlying CAN bus system on which NMEA 2000 is based, provides a very reliable and robust networking system with the ability to have priority messages and guarantees that messages will get through. Each end of a NMEA 2000 main network cable must be terminated with an approved terminating device to prevent signal reflection caused by a mismatch in impedance.

For specific one-way data traffic applications, NMEA 0183 remains popular due to its simplicity and the benefits of electrical isolation.

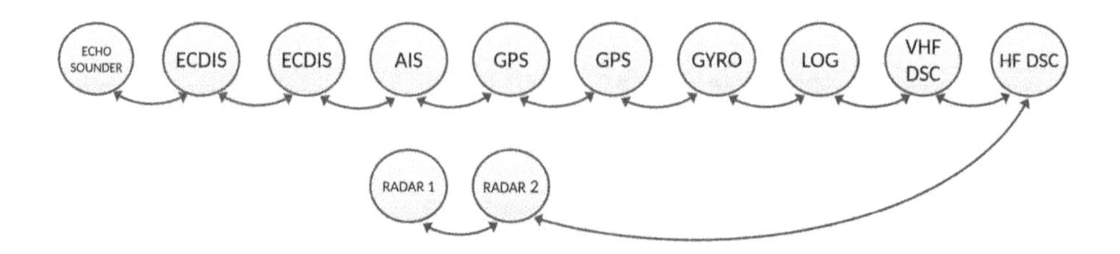

▲ **Figure 2.6** *Typical NMEA 2000 bridge wiring diagram*

Figure 2.6 shows a typical NMEA 2000 bridge wiring diagram. The two cable ends would be terminated as shown in Figure 2.5. Each NMEA 2000 device can be programmed to listen to the output of many sending devices that are connected.

Gyro Compass

When a ship is out of sight of land it is imperative for deck officers know what direction the ship is pointing. Even close to land, large vessels, manoeuvring slowly, need to know accurately the ship's heading. The ship's main gyro compass gives us the heading of the vessel with respect to true north. Gyro compasses use one of two methods to determine true north: a spinning mass or laser light.

Spinning mass gyroscope

A spinning mass 'wants' to point the same spot in space. In Figure 2.7 a spinning disk is shown. Its bearings are mounted in a metal frame called a gimbal, this gimbal is mounted within another gimbal whose bearing are at 90° to the inner gimbal. There may be three or four gimbals. The final, outer gimbal is fixed to the vessel such that when the vessel turns to port or starboard, the spinning disk is allowed, by the gimbals, to continue pointing to a fixed point in space. To make the spinning disk point to true north, as opposed to a fixed point in space, we rely upon a property of spinning disks called precession. Precession causes the spinning disk to turn to the left or right if we put vertical pressure on the end of the spinning gyro axis. In Figure 2.8, we can see the gyro disk turning in a clockwise direction (looking from above) if we press down at one end of the gyro axis. The gyro will precess in the other direction if we were to push up.

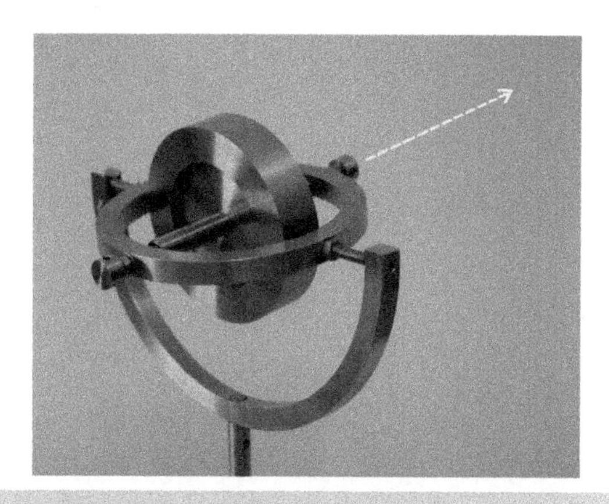

▲ **Figure 2.7** *Gyro pointing to fixed spot in space*

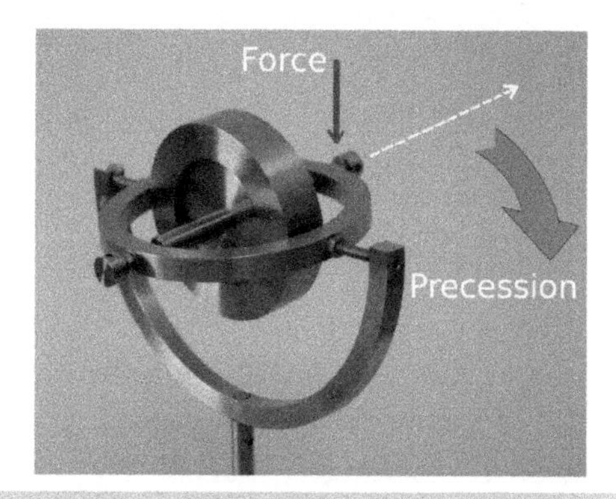

▲ **Figure 2.8** *Gyro precessing clockwise with downwards force applied*

Precession causes the gyro to point in different directions depending upon the force applied to either end of the axis of spin.

Assume the gyro is spinning and is susceptible to forces at either end of its axis. Assume the gyro is currently lined up north-south and is horizontal to the earth's surface. As the earth rotates, with the gyro always pointing to a fixed point in space, the spinning gyro would look from an observer on the earth's surface as though it was slowly pointing to different spots in space, 'wandering'. If we could sense the amount of vertical tilt, perhaps by using a pendulum which, under the influence of gravity, always wanted to point to the centre of the earth, we could apply pressure to the end of the gyro axis, which will cause the gyro to precess back to true north. If the pendulum was fixed to the body of the inner gimbal, each degree of tilt away from the earth's horizontal would increase the force acting at the end of the gyro spinning axis increasing precession causing the gyro to point to north.

▲ **Figure 2.9** *Navigat X Mk2 high performance gyro (front cover removed). © Sperry Marine*

As the earth rotates, the gyro 'wants' to tilt, causing the pendulum effect to precess the gyro back to north. The amount of tilt depends upon the latitude of the gyro. The distance from the equator affects how much 'correction force' the pendulum effect must apply to give the required amount of precession. To obtain accurate headings from the gyro, we need to input to the gyro an amount dependent upon the vessels latitude. Older gyros had a manual latitude input but all modern gyros have an electronic input, typically NMEA 0183, or other serial input from a GPS. Related to the latitude input is the speed input. The amount of force applied to the gyro to obtain the correct precession is dependent upon the speed of the vessel across the earth's surface combined with the heading of the vessel. To obtain the most accurate heading output from a spinning gyro, a speed input is used. Older gyros had manual inputs but modern examples use the log output or the speed output from the GPS. Figure 2.9 shows a modern high performance spinning mass gyro with the front cover removed (Sperry Marine Navigat X Mk 2).

Spinning mass gyros take some time to settle after power up. The forces that precess the gyro towards north are very small and need to be carefully controlled. Modern gyros are computer controlled and settle faster than earlier systems. To enable a gyro to settle more quickly, set the gyro heading to true north before power up. Typical settling times for modern, digitally controlled spinning mass gyros varies between 30 and 90 min.

Laser gyroscope

North seeking laser gyroscopes are relatively new. A laser gyroscope relies on the principle that the speed of light is constant. If you have light from a laser emitter split into two paths, one path travelling in a clockwise direction round a geometric shape such as a square or triangle, the other path travelling in an anti-clockwise direction around the same path, when the two beams of light strike a target simultaneously an interference pattern will be produced. If the speed of light is constant, when the geometric shape is rotated clockwise or anti-clockwise, one beam of light will be travelling further than the other, while the geometric shape is moving. This difference in speed of the two beams (with respect to the target) will cause an area of light and dark pulses to be seen and counted while the movement takes place. This count of light/dark pulses gives us an extremely accurate indication of how many degrees our geometric shape has moved. Laser gyros described so far make excellent accelerometers and angular movement indicators but they do not seek north. Non–north seeking laser gyros have 'aiding' inputs which allow them to be aligned with true north; once aligned they can measure heading changes very accurately. To make a laser gyro seek north, two more laser gyros need to be added to the system to help the heading gyro, gyros for pitch and for roll need to be added, enabling the movement of the vessel to be recorded in any direction along the X, Y and Z axis. An attitude/level sensor is fitted to detect gyro alignment with the earth's surface. All three aser gyros are aligned at 90° to each other and bolted to the hull of the vessel. If the three laser gyros were kept stationary with respect to space, no light/dark pulse transitions of light would occur. However, due to the earth's rotation, the three laser gyros detect the rotation of the earth, counting the light/dark pulse transitions from each gyro. The laser gyro is fed the latitude and longitude of the vessel, typically from a GPS. By using Kalman filtering, the gyro computer software is able to match the slow change in pulses from the three gyro outputs and signals from the attitude sensor caused by the rotation of the earth, and match them to the gyros location on the planet's surface. Only one course heading would match these inputs, enabling true north to be calculated. These slow changing signals are required to be extracted from the rapidly changing pulse counts caused by movement of the vessel.

Until Kalman filtering was developed in the 1960s, it was too difficult to effectively extract these slow and noisy 'earth movement' signals. Initially, Laser gyros where built with mirrors, with light travelling in straight lines, following a rectangular or triangular path. The manufacture of ring laser gyros (RLG) is too expensive for commercial marine use.

Fibre optic gyros (FOG) made laser gyros a commercial possibility. The transmission of laser light along fibre optic cable is commonplace and is used extensively in data communications. In one FOG compass configuration, laser light is split into two beams and sent through two separate coils of fibre optic cable, perhaps 500 m length. Each coil of cable is wound in opposite directions. The ends of the cable allow both laser light beams to merge on one target where patterns of light/dark pulses can be counted. FOG compasses are significantly more robust and cheaper that RLG compasses. Both types of laser compass are sealed for life. Figure 2.10 shows two coils of fibre optic cable, clockwise and counter clockwise, fed from a common laser source via a half- silvered mirror. Exiting from the coils, both beams of light impinge on a target where the two beams can 'beat' together, the beat frequency depends upon the rotation of the gyro assemble. The optical beats are detected by the light sensitive device.

All modern gyro compasses are computer controlled devices. On power up they perform built-in tests (BITs), issuing error codes or alarms when a fault is detected. Constant BITs are also used. Constant BITs or background BITs constantly check for the existence of correct inputs, validity of outputs and EPROM checksums. A Watchdog timer will cause the system to restart if a major hardware or software error occurs.

Figure 2.11 shows the Navigat 3000 FOG system, completely sealed for life, from Sperry Marine.

Figure 2.12 shows how important the gyro compass is to the safe navigation of a vessel. Its heading output is used by many items of equipment, including: steering repeater, autopilot, radar, AIS, ECDIS and many more.

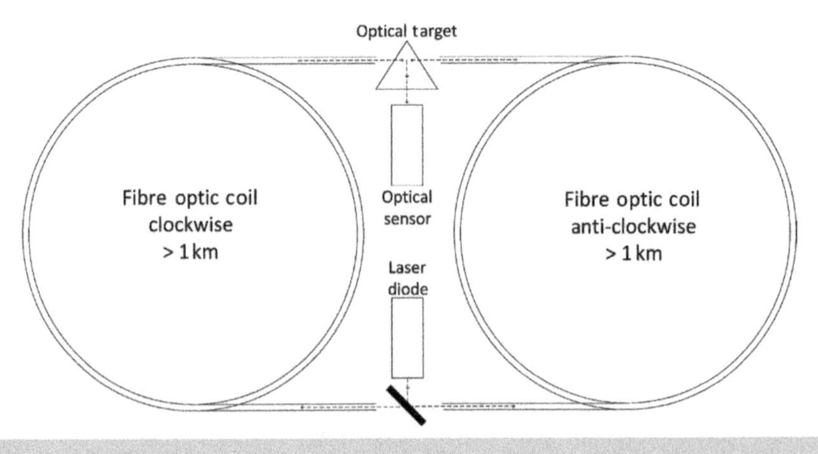

▲ **Figure 2.10** *FOG Gyro concept*

▲ **Figure 2.11** *Navigat 3000 fibre optic gyro. © Sperry Marine*

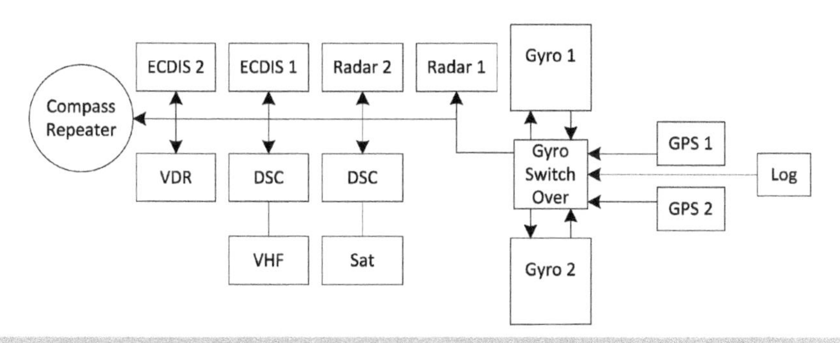

▲ **Figure 2.12** *Where the heading signal from the gyro is used*

Autopilot

Autopilots are fitted to comply with IMO regulations. Autopilots enable a vessel to follow a set course, adjusting the rudder angle to maintain the commanded heading. They save a great deal in manpower costs and steer a more accurate course than humans.

An autopilot effectively has two inputs (course required and current heading from a compass) and one output (signal to vessel's steering gear/rudder).

Figure 2.13 shows the front panel of a typical small-vessel autopilot. The large dial is the course required input. With this control, the deck officer can set the required course heading. The compass input is shown on the top left display. We do have the ability to choose from a variety of compass inputs. This is because our main compass may fail and the officer of the watch will need to quickly switch to another heading reference. Some ships carry multiple gyros, or a transmitting magnetic compass to serve as a fall back facility. Figure 2.14 shows an autopilot system suitable for larger vessels. This model has adaptive features enabling it to 'learn' the vessels steering characteristics.

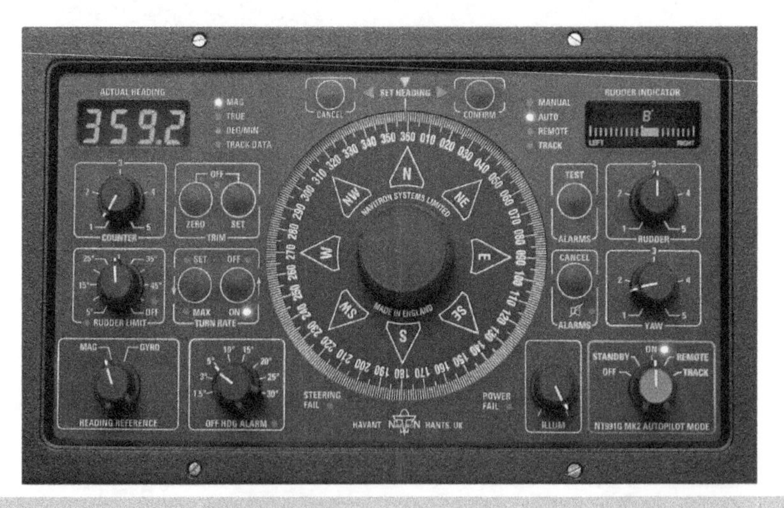

▲ **Figure 2.13** *Navitron NT991G Mk2 autopilot for smaller vessels. © Navitron Ltd*

▲ **Figure 2.14** *Navitron NT999G autopilot for large vessels. © Navitron Ltd*

The main output of the autopilot is a signal that commands the rudder to travel to port or starboard. The further off course the vessel is, the greater will be the commanded rudder angle.

For the autopilot to function in an effective and safe manner, the rudder angle needs to be measured and transmitted back to the autopilot. This allows the autopilot to monitor the rudder and its driving mechanism for effectiveness and failure. To allow the autopilot to be used in close quarter situations it is usual for bridge wings to be fitted with control units that allow temporary override of the autopilot. To ensure that the autopilot is working effectively and safely, the computer program controlling the autopilot constantly monitors its operations. The running of BIT at power up and constant BIT during operation is a constant theme of most professional equipment. We need equipment to be available constantly, and we need it to flag up problems as soon as they occur.

Figure 2.15 shows the internal functions of a simple autopilot. A modern autopilot suitable for installation on any vessel would have many more inputs and outputs that can be configured to interface with a variety of compasses and steering systems found in ships today. The output from the autopilot can be one of two types: follow-up (FU) and non-follow-up (NFU). FU is normal mode of operation. The autopilot requests a rudder movement, say 5° to port. The autopilot would energise the steering gear in the port direction until 5° to port is reached. The rudder has followed the autopilot request. It does not matter if the link between the autopilot and steering gear is analogue, digital or via relays, the rudder follows the autopilot command and ceases travel when the command is satisfied. It can be inconvenient to use the autopilot to make large course changes when navigating around obstructions. In these situations, the NFU mode can be beneficial. The NFU mode, when selected, allows the rudder to continue moving while a control on the bridge is operated. For example, the officer of the watch wants to

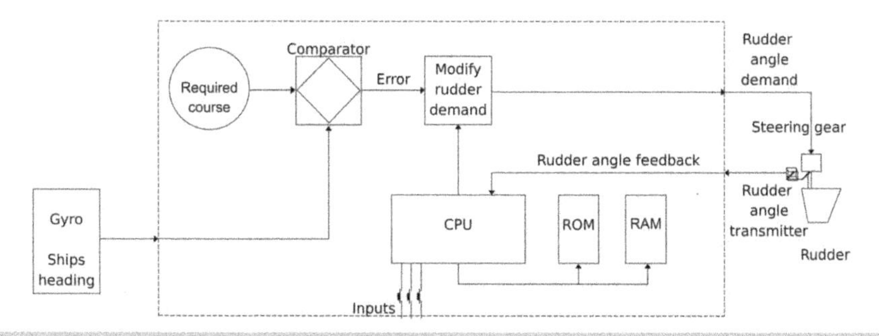

▲ **Figure 2.15** *Autopilot internals*

put the rudder to 20° to port, NFU mode is selected, the NFU control moved to port, the rudder starts moving to port, when the rudder reaches 20° to port, the officer releases the NFU control and the rudder remains at 20° to port until the control is operated again. It should be easy to switch between NFU and FU modes. An alternative to NFU mode is to switch to manual steering control.

For the autopilot to function correctly the rudder position must be transmitted to the autopilot. A rudder angle transmitter (RAT) is located in the steering flat linked to the tiller arm. As the rudder moves, the RAT linkage moves, transmitting the rudder angle to the autopilot and bridge rudder angle indicators (RAI). The RAT transmits either an analogue signal (0–5 V, 0–10 V, 4–20 mA) or a digital signal using NMEA 0183. Analogue RATs typically use a potentiometer to measure the angle, converting the position of the 'pot' to a voltage which is passed to the autopilot. Digital RATs typically use a rotary encoder to digitise the rudder angle. A small microcomputer within the transmitter converts this digital angle into a NMEA 0183 serial signal. Use of the NMEA 0183 is more popular now that voyage data recorders (VDR) are required to be fitted and need to store rudder angle data. In either case, the autopilot compares the rudder angle received from the transmitter with the requested rudder angle. If the rudder does not reach its requested angle within a few seconds the autopilot will trigger an alarm on the bridge to raise attention to the officer of the watch of a potential problem with the steering system.

Steering Gear

The steering gear converts the low voltage signals from the autopilot and helm into powerful signals sufficient to move the rudder. In large vessels, the size of rudders and the force acting upon them is such that hydraulic systems are used to power rudder movement. There are two main types of hydraulic systems: ram and rotary vane.

Hydraulic ram systems consist of a piston within a cylinder; hydraulic oil is forced under pressure into the cylinder pushing the piston. The piston is connected to the tiller arm which attaches to the rudder. The rotary vane system consists of a vane attached to the top of the rudder spindle, the vane being contained within a circular space. Hydraulic oil can be injected under pressure to either side of the vane causing the vane and rudder to move to port or starboard.

In each case hydraulic oil under high pressure is used to force movement of the rudder. The control of this oil pressure and flow enables us to move the rudder to any angle within its range. We are concerned with the control signals that cause the steering gear to move between port and starboard.

Steering commands from the bridge can be FU or NFU. Both types of signal can come from the main helm position, bridge wing steering positions and autopilot. In the distant past, steering commands have been transmitted by hydraulics, voltage or current. Many ships now use digital signals for this communications path. Whichever communications system is used, the steering signal arriving at the steering gear is required to operate an electric solenoid or torque motor to control the delivery of hydraulic fluid to the steering motor.

In Figure 2.16 we can see simplified examples of a ram steering system. A: and B: hydraulic cylinders, C: tiller, D: rudder angle link bar, E: rudder angle transmitter, F: rudder, G: electrically operated hydraulic valve from helm or autopilot, H: port and starboard rudder command signals, J: rudder angle signal.

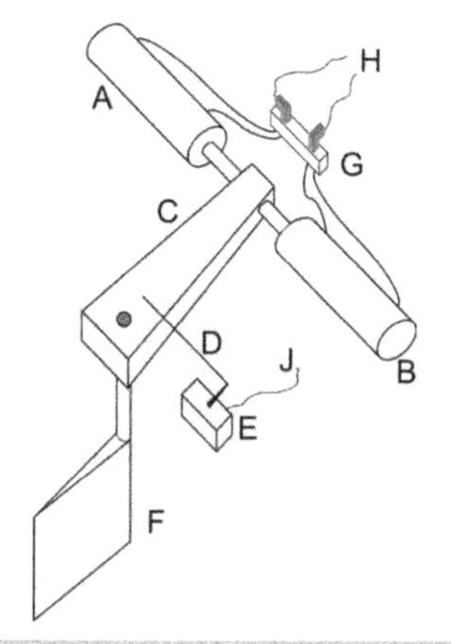

▲ **Figure 2.16** *Simple steering gear*

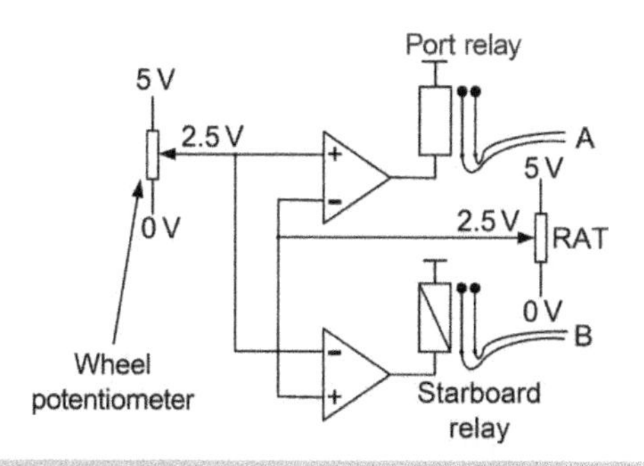

▲ **Figure 2.17** *Simple transmission of helm signals: follow up system*

In Figure 2.17 we can see simplified examples of steering command transmission systems. The wheel mounted potentiometer presents a variable voltage to two comparators. The rudder angle transmitter (RAT) presents a variable voltage to the same comparators. When the wheel is moved to 5° to port, the wheel voltage changes from 2.5 to 2.6 V. This causes one comparator to switch, causing its output to energise the attached relay, which causes the hydraulics to start moving the rudder towards 5°. Eventually the RAT will change from 2.5 to 2.6 V, at this time, causing the comparator to reset, the relay to de-energise and the hydraulics to lock the rudder in the new position. This implements a simple 'follow up' system.

The ship designer can select any type of steering motor (ram, vane or electric) with FU or NFU built into the steering motor, coupled with a wide choice of steering command transmission systems. Each ship is different and technology is making rapid advances in the area of steering control.

The rudder angle transmitter discussed above can be used to measure and transmit the rudder angle to the steering command system. In follow-up mode, the helm command and the current rudder angle are compared; any difference causes a signal to be sent to either the hydraulic solenoids or the swash plate of the variable delivery hydraulic pump causing the rudder to move to the required angle.

Echo Sounder

The depth of water under a vessel is extremely important. It is necessary to measure the depth electronically. Depth sounders or echo sounders work by bouncing sound waves off the sea bed. If we can estimate the speed of sound in water, and we measure the length of time it takes for a pulse of sound to travel from the vessel to the sea bed and back, we can calculate the depth of water under the keel.

Sound travels through sea water at approximately 1500 ms, depending upon temperature, depth and salinity. If we were to produce a short sound pulse and simultaneously start a timing system, when the feint echo returned, we could stop the timing system and use the time information to calculate the depth.

Figure 2.18 shows a block diagram of a simple echo sounder that can display the echoes returned from the sea bed. Echo sounder manufacturers use different sound frequencies but typical values are between 30 and 90 kHz. To enable the sound pulse to be transmitted into the sea water from the keel of the vessel, a strong transducer is

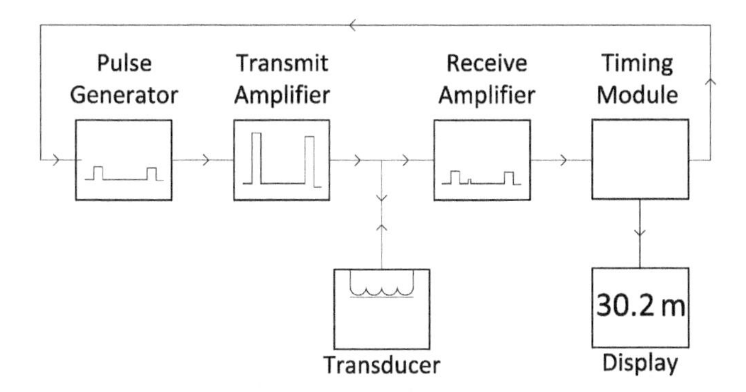

▲ **Figure 2.18** *Block diagram echo sounder*

required that can survive punishment from pounding. Typical transducers can be made from magnetostrictive or piezoelectric materials.

Figure 2.19 shows how a magnetostrictive transducer can be made. A coil of wire wound around a rod manufactured from a magnetostrictive material, is energised with a short length AC (alternating current) pulse; this pulse sets up a magnetic field that causes the rod to change length. By selecting the size of the rod to match the frequency of the pulse, large rod vibrations are coupled to the sea. When the AC pulse has ended, the vibration of the rod quickly reduces to zero. The rod now acts as a receiver. Feint sound echoes reaching the rod cause sympathetic vibrations in the rod, and magnetised rod movement causes small AC voltages to be induced into the coil of wire, the stronger the echo, the larger the voltage induced. This AC voltage is fed to the receiver. A realistic transducer would have many magnetostrictive rods to give a narrower and more defined beam width, similar in operation to a slotted wave guide for a radar antenna. Transducer frequency is matched to the transmitter and receiver frequency.

Display of returned echoes can be as simple as a numerical readout, 10.4 m in Figure 2.20, or a graphical display can be used, as shown in Figure 2.21.

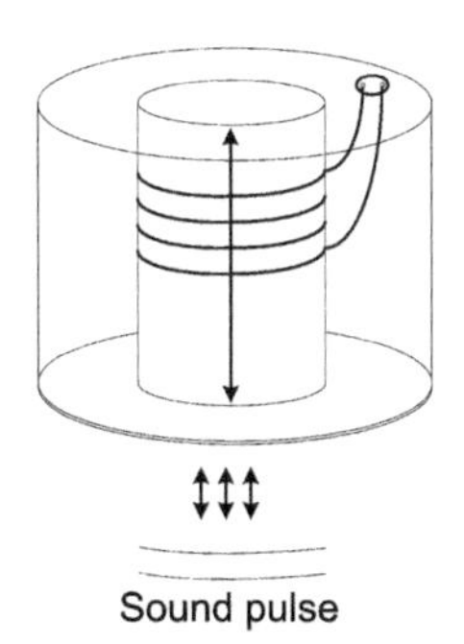

▲ **Figure 2.19** *Magnetostrictive transducer*

▲ **Figure 2.20** *Numerical depth display*

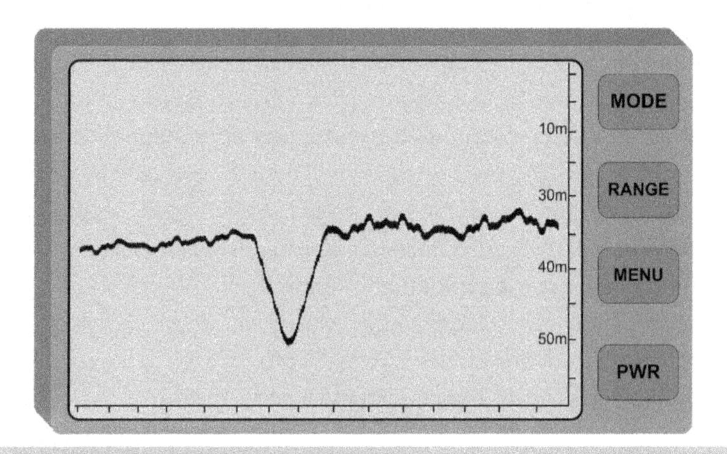

▲ **Figure 2.21** *Graphical depth display*

A numerical display simply counts up the microseconds between the transmit pulse and the first receive pulse then uses the formula:

$$\text{Depth} = \frac{v \times t}{2}, \text{ where } v = 1500mt = \text{time in seconds} \text{ to display the answer.}$$

Graphical display echo sounders measure the time the echo takes to make the journey to the sea bead and back but displays the returned echoes directly on the graphical display unit. This can be done typically in two ways: onto paper for a permanent copy or onto an LCD screen for an immediate indication only.

Paper based echo sounder displays, shown in Figure 2.22, consist of paper passing from spool to spool at a regular speed, controlled by an electric motor.

An electrical stylus presses onto the paper, the stylus moving from top to bottom of the paper. Echoes are amplified to a sufficiently high voltage and applied to the stylus. Whenever an echo is received, the high voltage on the stylus causes the paper to darken for the duration of the echo. As the paper is moving from right to left, a succession of marks are made on the paper. Each mark represents the depth of water at the time the mark was made. This gives us a permanent record of the sea bed the vessel has travelled over. Also, any echo received from objects between the vessel and the sea bed (such as fish) will also be recorded. Modern LCD systems replicate this paper-based style of functionality but without moving parts, and so are more reliable.

The transmitter, receiver and transducer are similar to the previous designs. For the received echoes to display on the LCD screen, a counter starts incrementing once per millisecond. Every echo received 'paints' a mark on the LCD screen at 'counts' pixels from the top of the display, so echoes from a deeper sea bed mark the display towards the bottom, shallow echoes mark the display towards the top. This gives the illusion that the display shows all echo returns from the keel of the vessel to the sea bed. The speed of the counter increment controls the maximum range and resolution displayed.

Many merchant vessel echo sounders are small units contained within one enclosure. The transducer is sealed for life, fitted to the keel of the vessel and inaccessible unless

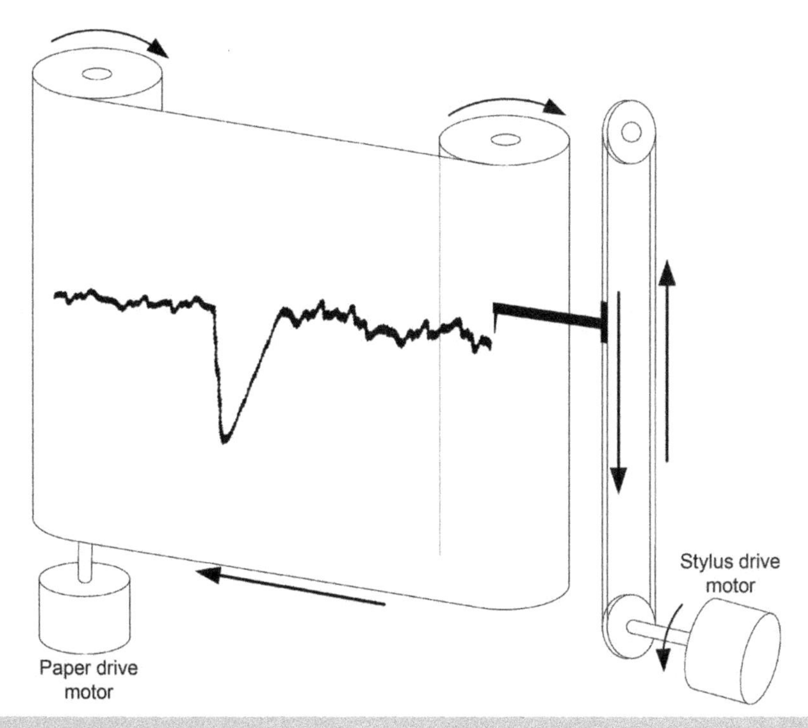

▲ **Figure 2.22** *Paper based depth display*

in dry dock. If an echo sounder is reported as faulty, we should check that it has power, that the cabling has the same resistance measurements found during installation and that the transmit pulse can be seen on the output terminals of the transmitter section.

Speed Log

The speed of a vessel can be measured using different methods. Speed through the water and speed across the ground can be significantly different.

Speed though the water on large vessels is measured using pressure differences with pitot probes, induced EMF and Doppler sonar. Pitot probes, like any protrusion from the hull of the vessel, are likely to be damaged in severe weather and are not generally used.

An EMF induced by moving a conductor through a magnetic field is used in the 'electromagnetic' speed log.

As shown in Figure 2.23, a coil of wire is energised with an AC voltage, causing a magnetic field to expand and collapse around the coil. The coil is surrounded by conductive water (any water that is not pure will conduct). When the vessel/coil move through the water, the sea water, acting as a conductor, has a small AC voltage induced. This small AC voltage is proportional to the speed difference between the coil and water. The faster the vessel moves the greater the voltage. As the voltage induced is small, it is usual to install an amplifier close to the speed sensor in the bilges to amplify the few hundred micro-volts from the sensor or convert it to a current or a digital signal prior to transmission to the bridge.

Figure 2.24 shows a typical speed log. The accuracy is affected by the salinity of sea water and its temperature. Most magnetic speed logs on merchant vessels will consist

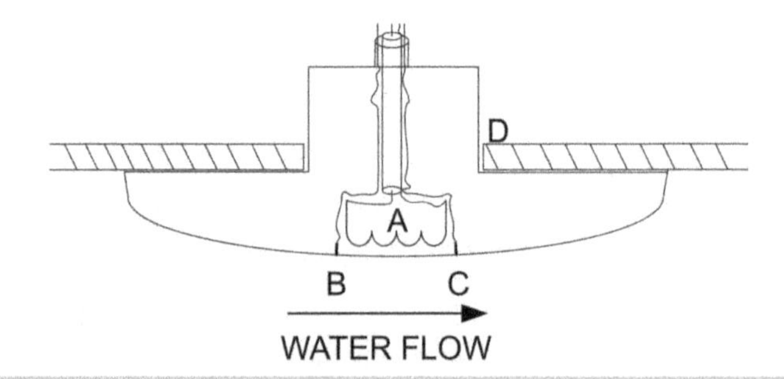

▲ **Figure 2.23** *Electromagnetic speed sensor*

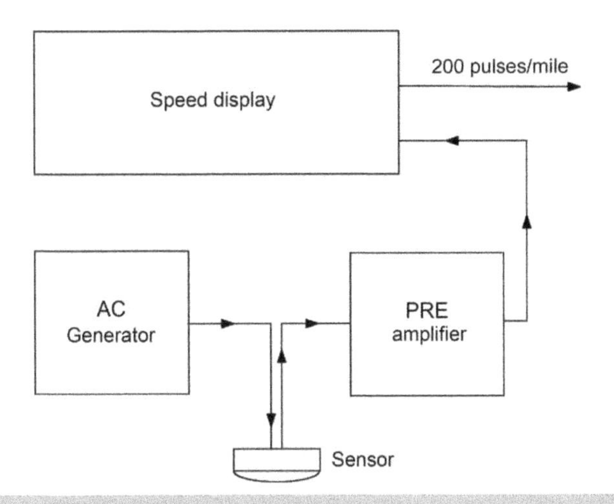

▲ **Figure 2.24** *Speed log block diagram*

of a display unit on the bridge with a preamplifier located adjacent to the sensor. Optional speed repeater indicators may be fitted on bridge wings.

Other navigational equipment may receive a speed signal from the log (Figure 2.25). A standard speed signal from a log is digital pulses sent at the rate of 200 pulses every nautical mile. These signals are transmitted to radar, ECDIS and VDR.

▲ **Figure 2.25** *Sperry Marine speed log*

Automatic Identification System

An automatic identification system (AIS) is fitted to all but the smallest vessels. It is a system that transmits vessel location, course and speed via a VHF radio channel to all ships within range (typically up to 30 miles). It also receives similar information from other equipped vessels within range. This enables the AIS system to display, on an electronic chart, all participating ships, within range, including their name, location, course and speed. As the display system is computer based, it can monitor the reported course of nearby vessels and raise an alarm if a vessel is predicted to approach your ship too closely.

An AIS system consists of the following parts: transmitter, two receivers, controller and display unit. The AIS system must constantly receive your own latitude, longitude, heading and speed. Some AIS units achieve this with a built-in GPS system, others rely on external devices to provide this information. When AIS is installed, the technician programs the ship's permanent details into the system. This includes the ship's 'Maritime Mobile Service Identity' (MMSI) number, which uniquely identifies the vessel and vessel length. Deck officers can program information that changes per voyage including destination, ETA and draft.

AIS gathers all of the required information, static and dynamic, constructs a sentence of information, then listens for an empty 'slot' to transmit in. There are over 2000 'slots' in each of two channels. Sophisticated software is constantly monitoring all slots on both channels to receive other ship AIS messages, and uses the information gained to decide which channel and which 'slot' to use.

For AIS to work successfully it needs to be supplied with appropriate variable inputs: heading, latitude and longitude and speed. The AIS will constantly monitor its inputs and raise an alarm if one of its inputs fails. Received AIS messages are now recorded on the vessels VDR but utilised by the ship's radars and ECDIS. AIS messages are usually transmitted to these other equipments by NMEA 0183 but other networking systems can be used. All equipment that makes use of AIS messages will be checking the quality of signal and lack or presence of a signal and will raise an alarm if the AIS messages fail to be received.

Figure 2.26 shows a block diagram of an AIS. Two receivers are constantly monitoring two channels (161.975 and 162.025 MHz). The controller builds up a list of timeslot activity for each channel. When ready, the controller selects the best transmit channel by using a complex computer program. The controller assembles heading, speed and location data, appends fixed data such as MMSI, calculates a checksum and sends the sentence. Any received information from other vessels is displayed on the screen.

This enables the ships shown in Figure 2.27 to be aware of each others presence; even ship A and ship B become aware of each other before they become visible around the headland. Ship C may be transmitting, but the symbol C is displaying indicates we have not received any data for 10–20 min. D indicates a shore station monitoring AIS information for use by authorities.

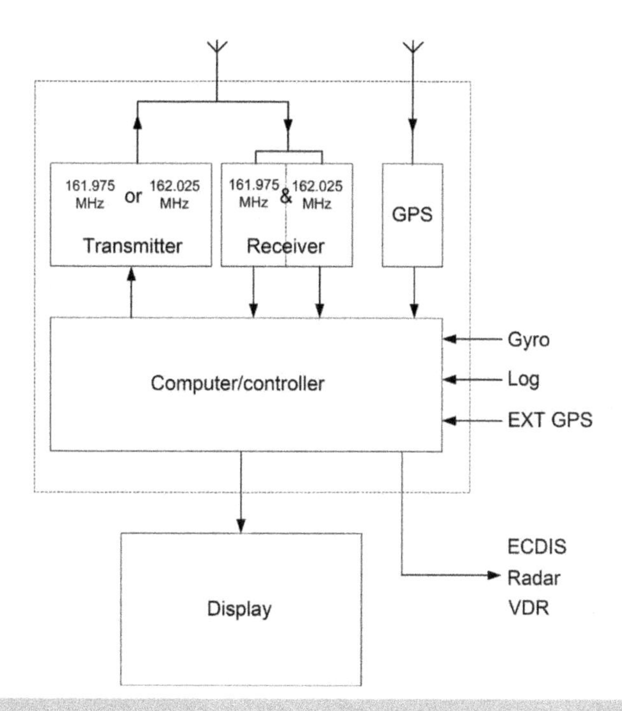

▲ **Figure 2.26** *AIS block diagram*

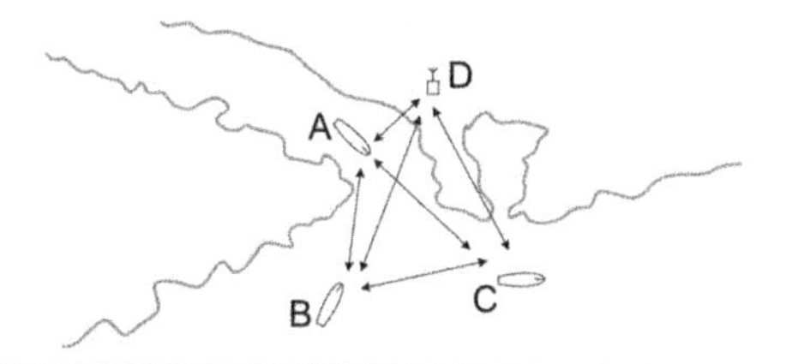

▲ **Figure 2.27** *AIS in use with ground station*

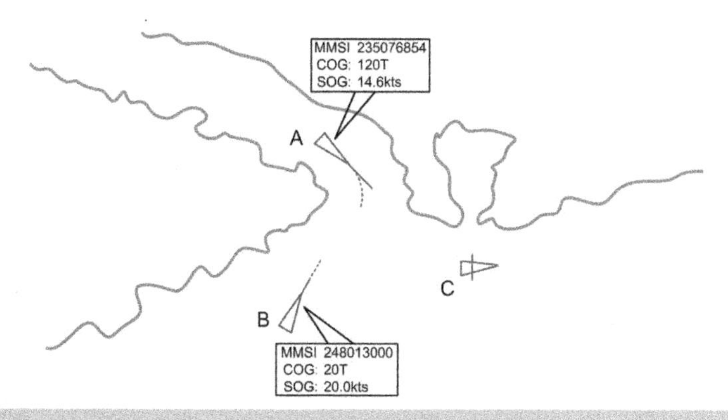

▲ **Figure 2.28** *AIS display*

Figure 2.28 shows the AIS display, with triangles for ships, dotted lines indicating which direction each vessel is travelling. Ship A indicates it is turning to starboard. Each active ship (moving and transmitting) can have an information box nearby indicated a variety of information that each other vessel has transmitted. In busy areas, these information boxes can be so numerous that they obscure the display, and hence are frequently disabled. A mouse click on any vessel causes key data to be displayed. It is normal now to display AIS data merged with radar data or ECDIS data. Communication of AIS data to other systems is usually by NMEA serial links.

With bridge equipment increasingly being networked together, Multi Function Displays (MFD) are more common. MFDs are Visual Display Units (VDU) or screens where the navigator can select what sensor information appears. This would allow you to select radar data, AIS data, ECDIS data to be displayed simultaneously. Care needs to be taken to avoid information overload. This allows the bridge team to be in a physically comfortable and practical location on the bridge and call up appropriate data that suits their current role. It allows easier fault finding and repair if a number of identical screens are used. It enables rapid selection of a new data source if say a radar fails, the other radar can be selected.

Long-Range Identification and Tracking

Long-range identification and tracking (LRIT), not to be confused with AIS, is a system that reports the vessel's identity (MMSI), position, date and time every 6 h to national data centres. This enables governments to track vessels approaching their coasts and shipping lanes. An approved, self-contained LRIT system consists of a stand-alone module that contains satellite communications and a GPS receiver. The GPS receiver is continuously receiving the vessel's position. Every 6 h, the satellite transmitter receiver sends a position report. It is necessary for the LRIT system to be able to send more frequent position reports if commanded by shore authorities.

If your vessel's main satellite terminal is suitable, the LRIT function may be programmed into your main satellite terminal to perform the LRIT function to avoid the necessity of purchasing a stand-alone system. However, if your main satellite terminal is shutdown in port, you lose the LRIT function and regulatory problems could arise. Figure 2.29 shows a typical stand-alone LRIT system (Sailor 6130 mini C LRIT).

LRIT data is confidential and is released only to qualified parties, unlike AIS data which is in the public domain and easily available to anyone.

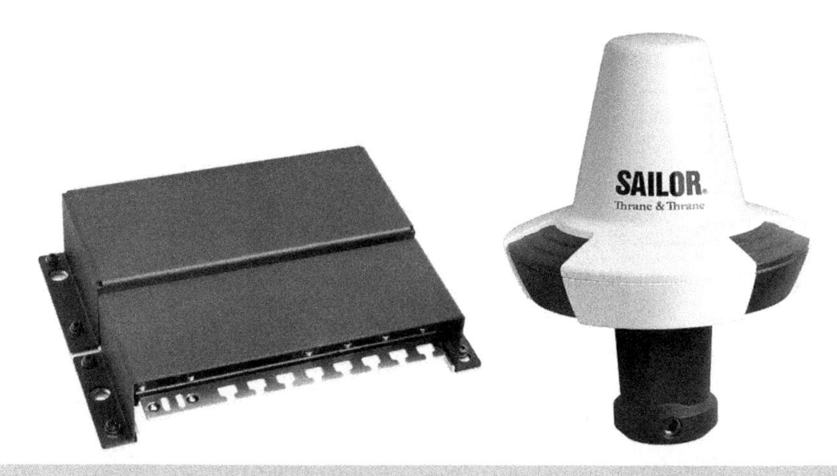

▲ **Figure 2.29** *Sailor 6130 LRIT self contained system.* © *Thrane & Thrane*

Global Positioning System

Global Positioning System (GPS) is a navigation system heavily relied upon by shipping. It consists of an array of satellites (constellation) which regularly transmit their precise position. A GPS receiver can use these transmissions to calculate its position.

As GPS satellites constantly orbit the earth, the accuracy of their orbital position changes due to local variations in the earth's gravity, space particles and solar radiation. Earth stations monitor each satellite producing accurate locations for each satellite. This location information is transmitted to each satellite frequently. Each satellite re-broadcasts this information. All GPS satellites transmit on the same frequency, 1.5 GHz, so complex coding schemes are used to enable receivers to differentiate between satellites. Satellites require extremely accurate time signals provided by atomic clocks that enable sub-nanosecond signal accuracy.

A GPS satellite's position in space is monitored by ground stations which update each of the 24 satellites on a daily basis. Each satellite transmits its orbital data every few minutes. This orbital data enables any receiver to calculate the satellite's precise position in space at any time. The satellite also transmits timing information repeatedly. This timing information, the accuracy being controlled by the atomic clocks, consists of a number of pulses. The receiver on board ship understands when this timing signal begins transmission. The receiver counts in nanoseconds how long it takes for the beginning of the timing signal to reach the receivers location. This process keeps repeating as the satellite passes over. If the receiver has a count of the precise time in nanoseconds for the signal to transit from satellite to receiver, we could draw a circle on a surface of the earth, centred on the satellites position at the time the signal began.

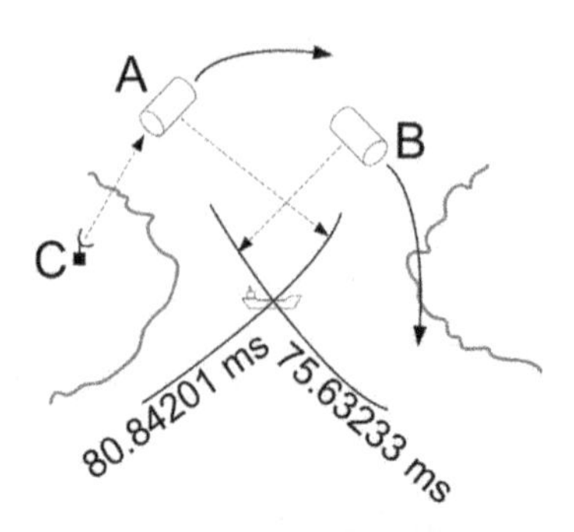

▲ **Figure 2.30** *GPS system*

If we repeated this process with another satellite, we could draw another circle on the earth, centred under this second satellite. This would give us two circles crossing, and two possible position fixes. If we repeated this with a third satellite, we would get just one position fix where all three circles cross (Figure 2.30).

Unfortunately the clock built into all GPS receivers is not accurate enough for this timing task. The flight time of the signal from satellite to receiver is between 60 and 100 milliseconds. During that time, the clock within our receiver may speed up or slow down. The second and third satellite signal may not be available for 5 or 10 min. During this time our receiver would be required to maintain an exceptionally accurate time. To overcome these timing constraints we receive signals from a fourth satellite. Using the fourth satellite enables the receiver to solve the problem of an inaccurate local time by allowing the solution of three simultaneous equations with three unknowns.

GPS can provide course and heading information as it is constantly calculating its current position. By reference to its position 1 min ago compared to the current position it can simply calculate its course and speed. GPS speed is Speed Over Ground (SOG) and course is true north. A block diagram of a GPS receiver is shown in Figure 2.31. The output of the receiver consists of ASCII characters, typically at 4800 baud over two serial wires using the NMEA 0183 protocol:

$SRRMC,219022.123,A,5084.6218,N,00130.0115,W,0.0,0.0,,,A*45

This shows a typical NMEA sentence, starting with a $ and ending with a carriage return and line feed. Parameters are separated by ',' and the * character terminates the content, followed by a two digit checksum. NMEA data can be displayed on a PC configured to show serial data.

GPS signals are affected by the ionosphere. This can give rise to navigational errors. If the satellites used to give a position fix are low on the horizon, the signal path the receiver 'sees' is longer, more electrically turbulent and more prone to variation, which will affect

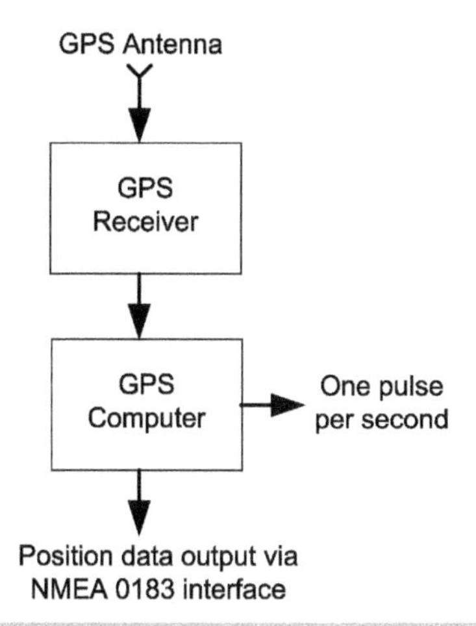

▲ **Figure 2.31** *GPS inputs and outputs*

the accuracy of the fix. Occasionally a GPS satellite can fail and it can take some minutes before the satellite is turned off. During this time all fixes using the faulty satellite would be suspect. GPS signals can be jammed. Due to the low power of the satellite signals, a local transmitter could be used to interrupt the use of GPS. The absolute accuracy of GPS cannot be guaranteed. If you are alongside and monitor the ship's position via the GPS receiver, you will see the ship apparently move slightly in a random way by a few metres throughout the day. The accuracy of your position varies with your latitude and longitude, the current satellites in view and levels of local interference. Large vessels have two GPS receivers, main and standby, to ensure availability.

Differential GPS

Due to occasional vagaries of the GPS signal, differential GPS (DGPS) was developed to provide a means to increase the accuracy of the GPS position fix. If a fixed land-based GPS receiver (reference station) plots its position, then, due to the GPS errors listed above, it appears that the fixed station is moving around within a circle of 30 m (occasionally larger). Since the fixed land-based receiver is not moving, this random position error is recorded as an offset to the actual position and sent via a radio transmitter to suitably equipped GPS receivers which can apply this error correction signal and remove the error from their own position fix. DGPS enables a position to be fixed to an accuracy of 10 cm. Figure 2.32 shows a DGPS system. Reference station D receives satellite positions, creates differences and transmits these differences to ships. DGPS correction messages are sent by beacon transmitters which have a range of up to 300 miles on frequencies near 300 kHz.

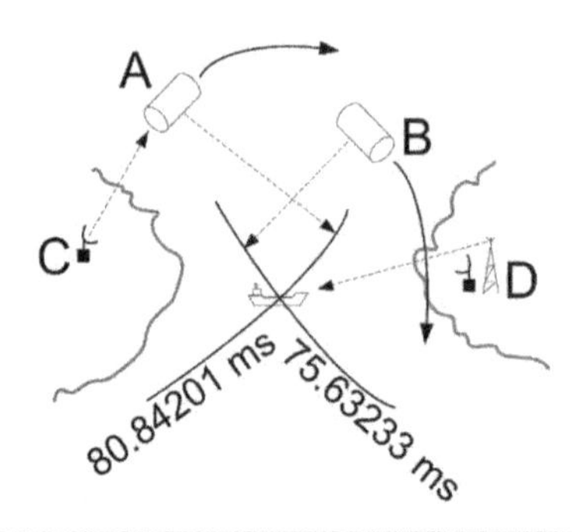

▲ **Figure 2.32** *DGPS system (additional ground reference station)*

Loran C

Loran C has effectively been phased out and is being replaced by an improved version called eLoran. To understand eLoran we will describe Loran C first.

Loran C is a terrestrial radio-based navigational system that consists of land based transmitters emitting signals on 100 kHz. Loran C transmitters work together in groups or chains. A typical chain consists of 3, 4 or 5 transmitters, separated by 500–800 miles. The 100 kHz signal propagates using a ground wave which ensures a very stable signal at each receiver. The power level of the transmitters is high, typically 250 kW, giving a range of over a 1000 miles from the centre of each transmitter chain.

If we recorded the time it takes for a pulse to travel from two transmitters, M and X, to a ship located at an equal distance from each transmitter, position 1 in Figure 2.33, we would see that the travel time of each pulse was the same, the difference in pulse arrival time would be zero. But the ship could be located anywhere on the line A–B and still measure zero difference in pulse arrival time. If the ship was located closer to transmitter M than transmitter X, at position 2, the pulse travel time recorded would be less for the pulse from transmitter M and longer from transmitter X, giving us a difference in pulse arrival times, but the ship could be located anywhere on line C–D

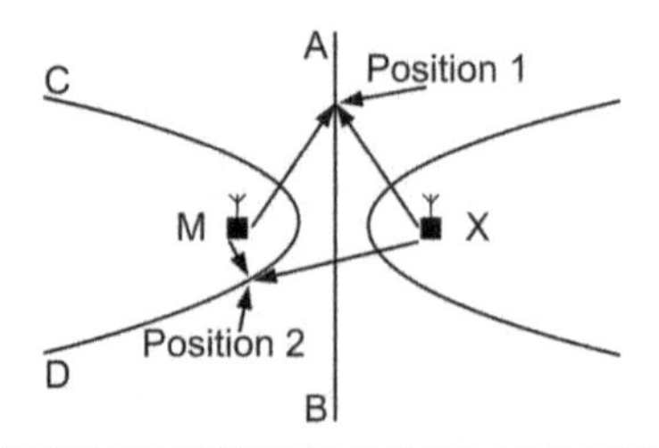

▲ **Figure 2.33** *Loran position line AB between stations M and X*

and still measure the same difference in pulse arrival times. To fix the ship's position, we need two position lines. A third transmitter is added, Y, shown in Figure 2.34.

We can now measure the difference between pulse arrival times arriving from transmitters M–X, M–Y. With 3 transmitters sending pulses simultaneously, we could measure the difference in pulse arrival time, and find one location where these 'times of difference' coincide. This is circled in Figure 2.35.

There will be only one location where the measured difference in pulse arrival times from two transmitter pairs M–Y and M–X will be found.

Figure 2.36 shows the geographic layout of a European Loran C chain.

If the pulses were to be transmitted simultaneously, our receiver, while able to time the different arrival times would not be able to differentiate which pulse came from which transmitter. Loran C adopts the following approach to remove the problem of transmitter/slave identification. The master transmitter M transmits a burst of eight, 200 μS pulses each 1000 μS apart, with a ninth 200 μS pulse separated by 2000 μS. Slave transmitters do not transmit a ninth pulse. When a slave transmitting station receives the pulses from the master station, it waits a predetermined time then transmits its eight slave pulses.

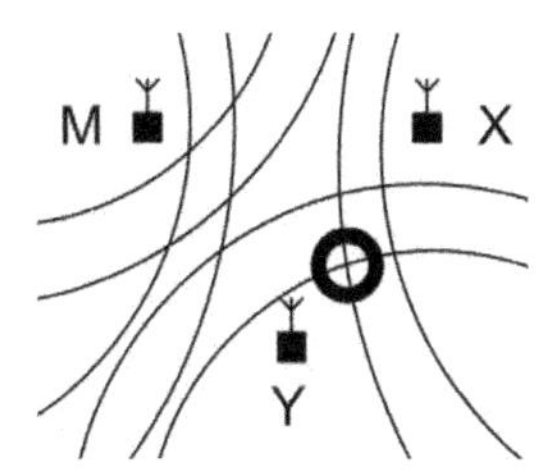

▲ **Figure 2.34** *Loran with 3 transmitters, multiple lines crossing*

▲ **Figure 2.35** *Loran master with 3 slave stations*

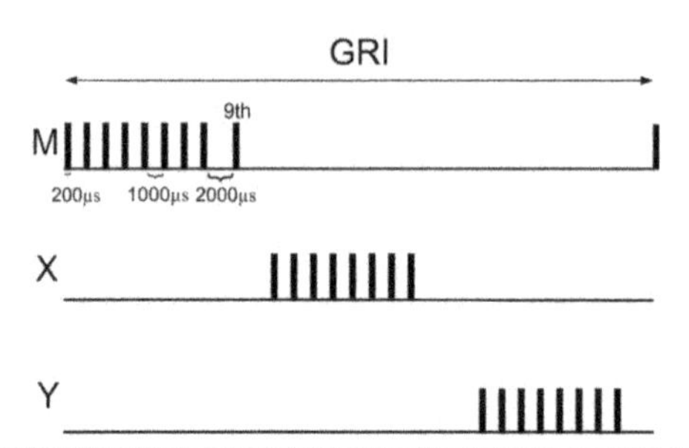

▲ **Figure 2.36** *Loran signals from master with 2 slaves*

The next slave in the chain waits a predetermined delay then transmits its eight slave pulses. Using this method, a receiver located anywhere in the service area of a Loran C transmitting chain is guaranteed to receive groups of pulses from each transmitter sequentially with no possibility of the master or slave transmitters interfering with each other in the same chain. Loran C chains are identified by the total transmit cycle time.

Figure 2.36 shows the pulse trains for a two slave station Loran C system.

The speed of radio waves is assumed to be 3×10^{-8} m/s; unfortunately, this speed varies depending upon the medium through which the radio wave passes: vacuum; air; passing over land or sea; all affect the speed. As we are measuring time differences, we need to take this speed variation into account. This is done by using a system called phase factors, secondary factors and additional secondary factors (ASFs), which are adjustments to timing to make allowance for the different types of surface the radio waves pass over to reach the receiver. ASFs are a lookup table which contain timing differences between the pure timing of the radio waves travelling at 3×10^{-8} m/s and the timing expected taking into account the nature of the surface between the transmitters and receiver at each location within the service area of the chain. Loran C gives position accuracies within 460 m.

eLoran

Due to budgetary cut backs in the United States, LORAN C transmitters have ceased transmitting. After an evaluation of GPS reliability, security and robustness, the United States declared an interest in the eLoran system to complement GPS navigation and to act as a fall back system. The United States is currently evaluating eLoran.

General Lighthouse Authorities have been conducting trials of GPS receivers on board vessels at sea. It was found that a low power GPS jammer could stop all GPS receivers from working correctly within a 20 mile area. Types of errors caused included: incorrect speeds, incorrect positions, receivers shutting down which triggered alarms throughout the electronic aids to navigation that made use of GPS outputs such as speed and position.

In Europe, the eLoran system has developed as the 'Eurofix' system. International standard ITU-R,M.589–3 documents data transmission using the eLoran 100 kHz signal to send differential corrections to both eLoran and GPS receivers. The difference between Loran C and eLoran include: increased timing accuracy of transmit pulses by using multiple atomic clocks at each transmitter site, monitoring pulse transmission accuracy by reference stations, solid state transmitters giving greater reliability and the transmission of additional correction data via an ELoran data channel embedded within the eLoran signal. Figure 2.37 shows a combined eLoran and GPS receiver from Reelektronika.

An eLoran receiver is a modern digital signal processing device that listens to 'all in view' eLoran stations. An eLoran data channel allows the sending of updated ASF information for eLoran and error corrections for GPS. The combination of 'all in view' reception, enhanced transmitter timing accuracies and the eLoran data channel improve eLoran position fixing accuracies to 10–20 m. The eLoran data channel uses eLoran pulses 5 through to 8, moving their transmission time by + or −1 μs. This pulse movement or modulation enables the eLoran system to transmit additional position correction information that can be used by GPS and eLoran receivers. Any clock problem or propagation error that could cause an incorrect position to be output by a receiver is quickly notified to users. Ships should minimise navigational risks by using complementary position fixing systems. An eLoran receiver complements the GPS, giving a robust position fix. Figure 2.38 shows the simplicity of a modern 'all in view' eLoran receiver.

▲ **Figure 2.37** *eLoran and GPS receiver from Reelektronika. © REELEKTRONIKA b.v.*

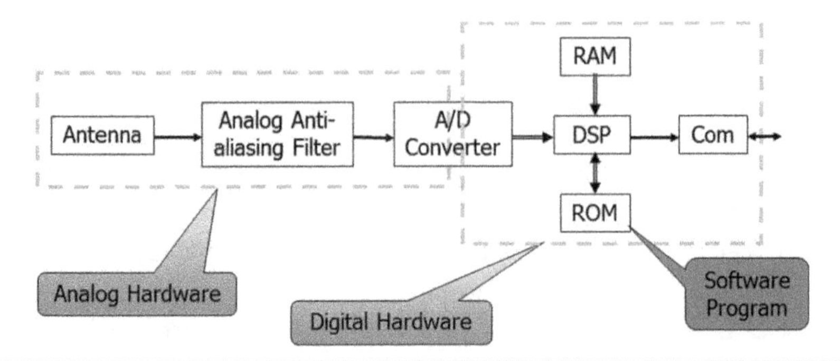

▲ **Figure 2.38** *eLoran block diagram. © REELEKTRONIKA b.v.*

Radar

Radar is an object detection system that bounces radio waves off objects, displaying the received echoes.

A radar system consists of a high power, high frequency transmitter that can transmit very short pulses; a sensitive, high frequency receiver able to receive very weak echoes from distant objects. Figure 2.39 shows a typical block diagram of a conventional basic radar.

Antenna

To be useful the radar needs to transmit a narrow beam of radio energy. Antenna design rules dictate that to achieve a narrow beam width, the antenna needs to be

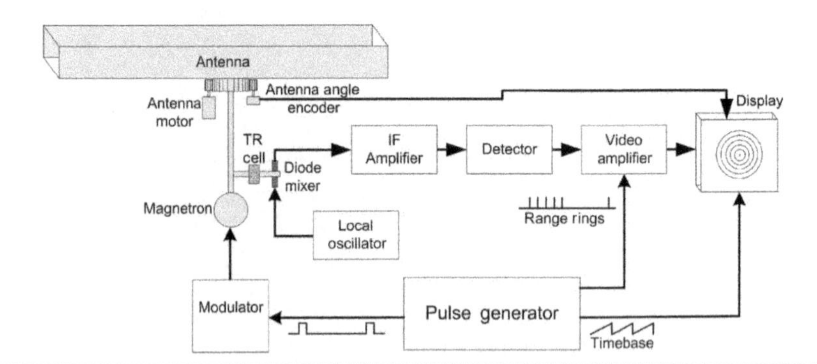

▲ **Figure 2.39** *Radar: Basic block diagram*

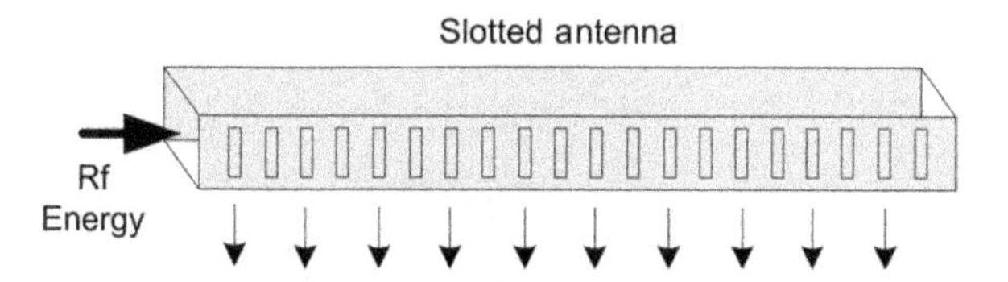

▲ **Figure 2.40** *Radar: Slotted antenna*

very large or the frequency needs to be very high (Figure 2.40). A practical size limit for ship borne antennas is approximately 2 m. This leads us toward a transmitter frequency of greater than 3000 MHz. The radio wave pulse is fed to the antenna. If slots are cut into the wave guide wall, the radio waves will emerge from these slots. If these slots are sized to fit the frequency in use, the slots act as antennas. Each additional slot causes the beam width of the transmitted signal to become narrower. With a narrow beam width, which gives good bearing discrimination, we get a noticeable squint. This is the radar energy effectively leaving the antenna not perpendicular to the front face but a few degrees off. Squint has to be accounted for when the radar system is installed on a vessel otherwise all bearings indicated by the system will be off by ~5°. On modern radars squint can be allowed for using installation menus.

Magnetron

When radar was invented it was difficult to generate radio frequencies of 3000 MHz until the Magnetron was invented. This device allowed the easy creation of 3000 MHz at high power. It typically consists of solid copper block with slots or circular chambers machined into it. In the centre a cathode is heated by a small voltage (Figure 2.41).

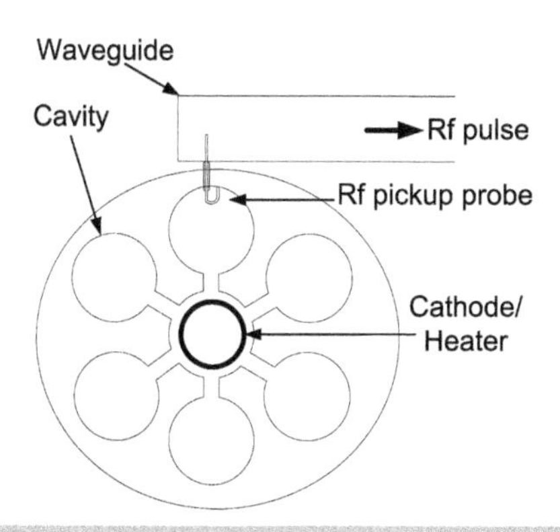

▲ **Figure 2.41** *Radar: Magnetron internals*

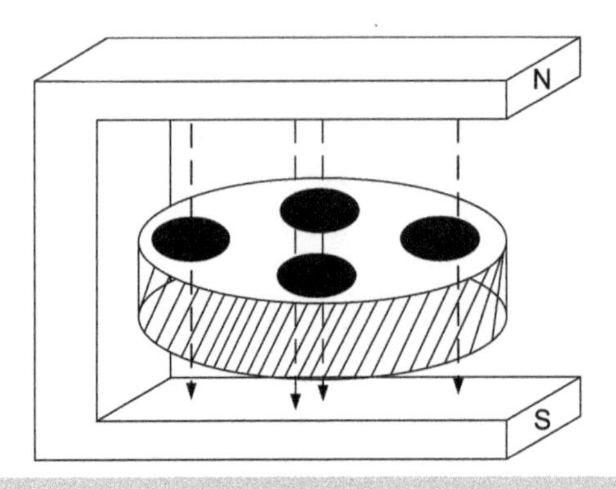

▲ **Figure 2.42** *Radar: Magnetron surrounded by magnet*

The copper body has a 2 kV pulse applied. If the cavities maintain a vacuum, then the voltage differential between the cathode and the body (2 kV) causes a stream of electrons to move between the cathode and body. If a strong magnet bathes the copper body with a strong magnetic field, electrons leaving the cathode follow a curved path, sweeping past the entrance to the slots or cylinders. As the electrons sweep past the entrance of each cavity, an electrical wave sweeps into the cavity, setting up oscillating waves, the frequency of oscillation is determined by the size of the machined cavities (Figure 2.42).

While the 2 kV difference exists, high power high frequency radio energy will be available from the magnetron. Radio waves are extracted from the magnetron by a short metal dipole inserted into one of the machined cylinders. This is the only exit for the radio waves. These radio waves are led to the antenna via a wave guide.

To enable good discrimination between targets that are on the same range from the ship but close to each other in azimuth, we need a narrow beam width from the scanner, otherwise the radar will start to receive the next echo before the first echo has stopped reflecting. This is shown in Figure 2.43(a).

To enable good discrimination between targets that are on the same bearing from the ship but close to each other, we need a short transmit pulse length, typically between 0.1 and 4 μs. This is due to the echo received from the near target still being received when we start to receive the second target's echo. The display shows both targets merged together. This is shown in Figure 2.43(b).

Modulator

Generation of a 1μs 2kV pulse is undertaken within the Modulator. Logic level pulses are created in the timing generator, passed to the Modulator, Figure 2.44, where Tr1

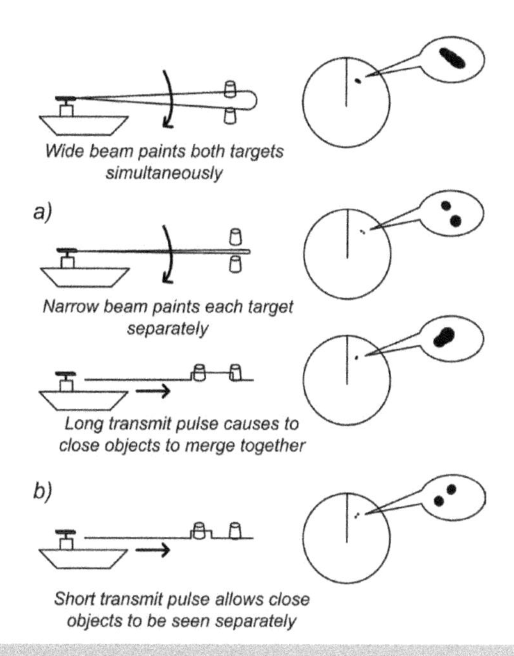

▲ **Figure 2.43** *Radar: Beamwidth and pulse length conflict*

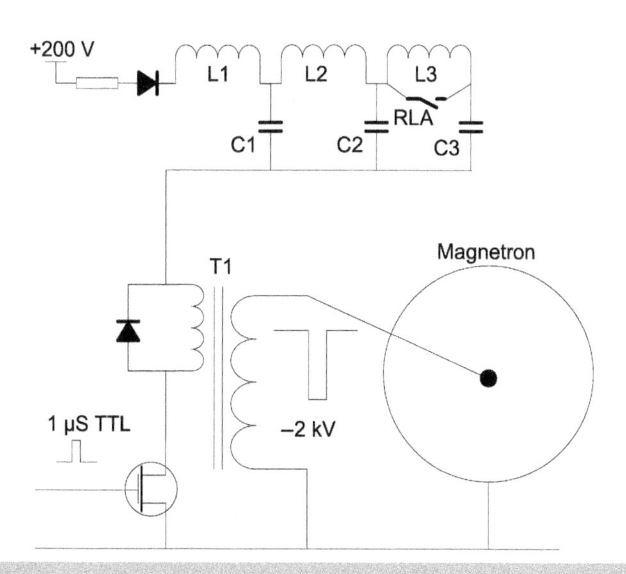

▲ **Figure 2.44** *Radar: Basic modulator circuit*

is turned on for 1μs. The rapid decay of 300 V to 0 V causes a high power pulse to enter the pulse forming network (PFN). The combination of L1 with C1 and L2 with C2 etc. gives rise to a HV 1 μs pulse fed into the pulse transformer. T1 matches the low impedance of the PFN to the Magnetron cathode, stepping up the voltage to 2000 V.

The Magnetron body is connected to earth, which makes maintenance operations safer. The 2000 V pulse connected to the cathode therefore needs to be negative (–2000 V). To obtain different pulse widths, different sections of the PFN are switched in or out using relay contacts. When RLA is energised, L3 is shorted out causing the generated pulse to reduce by approximately one-third. Although this modulator circuit is crude, it does give control of the pulse width and is simple and reliable.

TR cell

The weak signal (echo) returned from an object is very weak and at the same frequency of the transmit pulse. The receiver requires a very high gain to boost the echoes to a sufficient level that they can be displayed on the radar display. The high power from the Magnetron and the extreme sensitivity of the receiver require excellent isolation between the transmitter output and the receiver input. To achieve this level of separation we adopt a number of techniques: circulator/isolator, and transmit/receiver cell (TR cell). The circulator/isolator is a simple device that allows radio waves to travel through a ferrite device in one direction but not backwards by relying on magnetic fields and the properties of ferrite material. The TR cell, sometimes used without a circulator, is a device with two radio frequency (RF) windows. The transmitter pulse enters the TR cell and ionises the gas inside. When the gas is ionised it becomes a short circuit to the high power radar pulse, reflecting the pulse back to the antenna. When the transmit pulse ends, after say 1 μs, the ionisation of the gas stops and the TR cell becomes transparent to RF. To ensure the TR cell ionises quickly, a small radioactive source is placed inside the TR cell, with a level of activity such that the cell does not ionise without any transmitter RF. When TR cells age they start to allow some of the transmit energy through to the diode mixer, eventually causing the mixer diode to fail.

Local oscillator

Due to the difficulties of amplifying frequencies this high, we use the superheterodyne principle to convert the 3000 MHz frequency to 100 MHz or less. To convert 3000 to 100 MHz we feed a 3100 MHz into the mixer diode. The mixer diode sits within the wave guide, and is sensitive to the 3000 MHz frequencies. We create a 3100 MHz local signal in a local oscillator (LO). The two frequencies (LO at 3100 MHz) and the echoes at 3000 MHz mix within the junction of the semiconductor mixing diode. The outputs of the mixer diode are the sum and difference products: 3000, 3100, 6100 and 100 MHz The 100 MHz frequency is selected by filters tuned to 100 MHz and amplified within the intermediate frequency (IF) amplifier. The LO is usually a Gunn diode mounted within a cavity tuned to 3100 MHz. A Gunn diode is unique in that it enables oscillations when biased with a critical voltage. The frequency of the Gunn diode LO can be fine-tuned by applying a variable voltage to a Varactor diode within the cavity. This allows fine tuning of the LO to match the transmit frequency of the Magnetron plus 100 MHz.

Intermediate frequency amplifier

The 100 MHz frequency output from the mixer is very weak, the IF amplifier boosts the weak signal up to levels that allow normal radio processing methods to be used. The output of the IF amplifier are 100 MHz radio waves whenever an echo is received. To obtain the original shape of the echoes, a diode rectifier is used to demodulate the 100 MHz echoes into DC (direct current) pulses. The shape of the pulse is directly related to echoes received by the antenna. These DC pulses are amplified into levels suitable for use on a simple display screen. Each echo brightens the display, relating the object's position relative to the radar antenna bearing and the distance of the target and distance from the vessel..

Timing and control system

The timing and control system is in overall control of the radar system. Magnetron pulses from the modulator are triggered from here. The transmitter pulse width, pulse repetition period, range, range rings, variable range marker (VRM), variable bearing marker (VBM) and display controls are implemented here. Figure 2.45 shows the relationship between the major signals used within the radar system.

Figure 2.45(a) shows the transmit pulse, 1 μs wide sent every 1000 μs so the pulse repetition rate is 1000 μs. The average power of the transmitter is 1/1000 of the peak power. Typically, the transmitter power is 12.5 kW peak and 12.5 W average. The echoes can be seen in Figure 2.45(b). Sea clutter swept gain is shown in Figure 2.45(c) with the gain reduced early on in each transmit cycle, reducing the effect of echoes from nearby waves obscuring nearby targets. Figure 2.45(d) shows range rings. These concentric rings, typically six rings per display range, give an accurate indication of distance on the display. Typically, on a 24 mile range, the range rings will be 4 miles apart. To complement the fixed range rings, a VRM is arranged to give one pulse per transmit cycle, giving one ring on the display. By altering when this pulse occurs, the distance of the VRM from your vessel can be changed. The precise distance from your vessel to

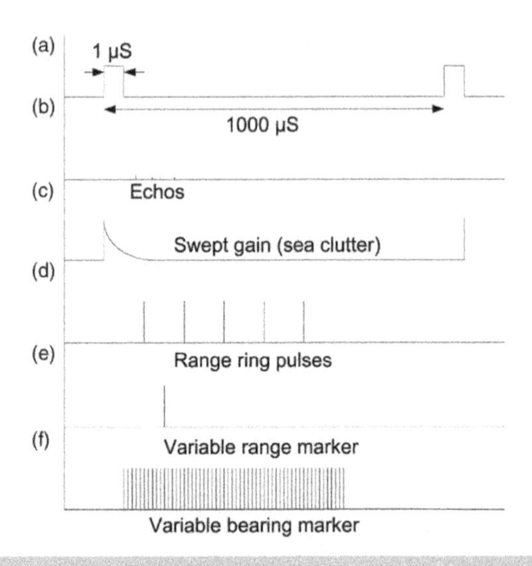

▲ **Figure 2.45** *Radar: Major internal signals*

range ring is displayed. Figure 2.45(f) shows the electronic bearing marker (EBM). This is generated once per antenna revolution, painting a dotted line on the display from your vessel to the circumference. By adjusting when during the rotation of the antenna the EBM occurs, you can change the bearing of the dotted line on the display.

The speed of rotation of the antenna should be constant, typically 24 RPM. The faster the antenna rotates, the lesser the potential number of echoes received from each object. The slower the antenna rotates, the longer it will take to update the radar display. This can be a problem when the vessel is manoeuvring. Due to wind loading, the absolute speed of rotation will change. The radar display needs to be in synchronism with the antenna. To achieve synchronism, the antenna heading is continuously sent to the display. A number of methods are used for this task. Two popular methods are syncro transmitter/receiver and logic pulses.

A syncro transmitter can be considered to be a rotating three phase transformer. The three secondary windings at 120° to each other are fixed; the primary winding is connected via a gear train to the antenna. As the antenna rotates, the primary winding, excited by a 400 Hz sine wave rotates. The rotating field is picked up by the three phase secondary windings. The voltage induced into each secondary winding depends on the angle of the primary with respect to the secondary. The syncro receiver is of identical construction, with three winding at 120° to each other and a rotating primary. The three phase windings of the transmitter are connected to the three phase windings of the receiver. The rotating field within the transmitter is transferred to the receiver.

The signal induced into the rotating receiver is amplified and fed to the control system which adjusts the speed of the receiver rotation. In this manner the rotating receiver is kept in synchronisation with the transmitter. Unfortunately because the rotation of the transmitter is geared up from the antenna, after system power up, the receiver and transmitter need to be aligned.

The pulse transmission system simply sends typically 10 bits of data during each degree of antenna rotation. When the antenna is aligned with the ships centre line, a pulse is sent (Heading Line) to the display ensuring synchronism. After power up, after one complete rotation of the antenna, the receiver and antenna will be in synchronism.

Sea clutter

Sea clutter is unwanted radar returns (echoes) from nearby waves. Due to the angle of the front of each wave near the vessel, large waves give large echoes. This can obscure navigational echoes. To assist in the partial removal of these unwanted returns, we need to reduce the gain of the receiver only for the first few miles of the radar range. This means we need to adjust the gain for every radar pulse transmitted. As the transmit pulse travels away from the ship, and echoes begin to return, we need to sweep the receiver gain from low to high, this is sometimes called swept gain. This is shown in Figure 2.45(c).

Rain clutter

Rain can reflect radar energy which sometimes covers the display with unwanted returns obscuring potential objects. As rain clutter can occur anywhere on the display, a different approach is required. Rain clutter can be partially removed by differentiating the signal prior to displaying it. The differentiated signal retains its edges but loses the bulk of the echo. This can help show objects obscured by rain.

North Up

With the radar operating in heading up mode, the top centre of the display represents the vessels heading, with the heading line at the top indicating the vessel's course. The radar display shows what could be seen through the bridge windows. As all navigational charts are shown with True north at the top of the chart, the radar display can be set to display 'North Up' to match the chart. To enable 'North Up' the vessel heading (course) from the gyro compass is required to be input into the radar display. The gyro heading causes the radar display to rotate until North is at the top of the display. The heading line now shows the ship's compass heading. Using 'North Up' enables the deck officer to view land objects on the display and match them with the chart. Gyro heading data can be transmitted by the syncro transmitter/receiver system or by serial data. Modern radar uses NMEA 0183 serial data (Figure 2.47).

Performance monitor

It is difficult to test the performance of a radar system. Two effective ways of testing the performance of a radar system are in use: an echo box and an active transponder.

The echo box is a small box which resonates at radar frequencies when enabled. To enable an echo box (performance monitor) a button is pressed and held near the radar. This enables an electrical supply to the echo box, causing the size of a tuned cavity to alter, rapidly sweeping through the possible radar frequencies in use. When a transmit pulse encounters the cavity, the pulse echoes around the cavity and a strong echo is sent to the antenna. This strong echo causes a large 'plume' to be shown display. The length of the plume is determined by the power output from the Magnetron and the gain of the receiver. If the length of the 'plume' is recorded during radar installation or after major servicing, slow degradation of performance can be seen. If the Magnetron is reaching its end of life and its power output is reducing, the 'plume' will be shorter. If the mixer diode or TR cell are ageing then this will also result in a shorter plume.

An active transponder performance monitor is a refined version of the echo box. Built into the casing of the radar turning gear, a small independent, calibrated receiver picks up the transmit pulses, measures the strength of each pulse, then transmits a very weak signal which the antenna picks up. The weak signal transmitted is encoded to indicate the effectiveness of the transmit side of the radar system. Typically the encoded signal shows up on the display as a set of arcs, centred on the vessels location. The circumference of the arc in degrees is a direct indication of the amount of power contained within each transmitted pulse. This gives a direct reading of the Magnetron, wave guide and antenna efficiency. The number of arcs (similar to the length of the plume) gives an indication of receiver sensitivity/gain (Figure 2.46).

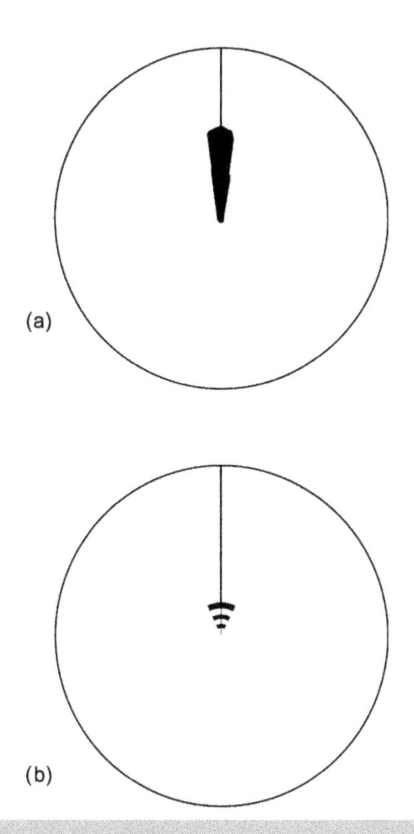

▲ **Figure 2.46** *Radar: Performance monitors*

True motion

If we were to set the radar display to True North and input our speed of the ground (SOG) into the radar display, the radar display would show our vessel 'moving' across the display in the correct direction of travel and all ground objects would appear stationary. True motion, when enabled, relocates the vessel's position on the display away from the direction of travel, giving a much longer display range. The movement of the vessel across the display follows the heading line at the speed input from the log or GPS. When the vessel's position on the display has reached within 30% of maximum travel, the system will relocate the vessel position back to 30% from the opposite side of the screen edge. True motion can give the navigating office a better indication of the situation, but concern can occur when the true motion system resets at the end of travel and a few second pass before the screen stabilises (Figure 2.47).

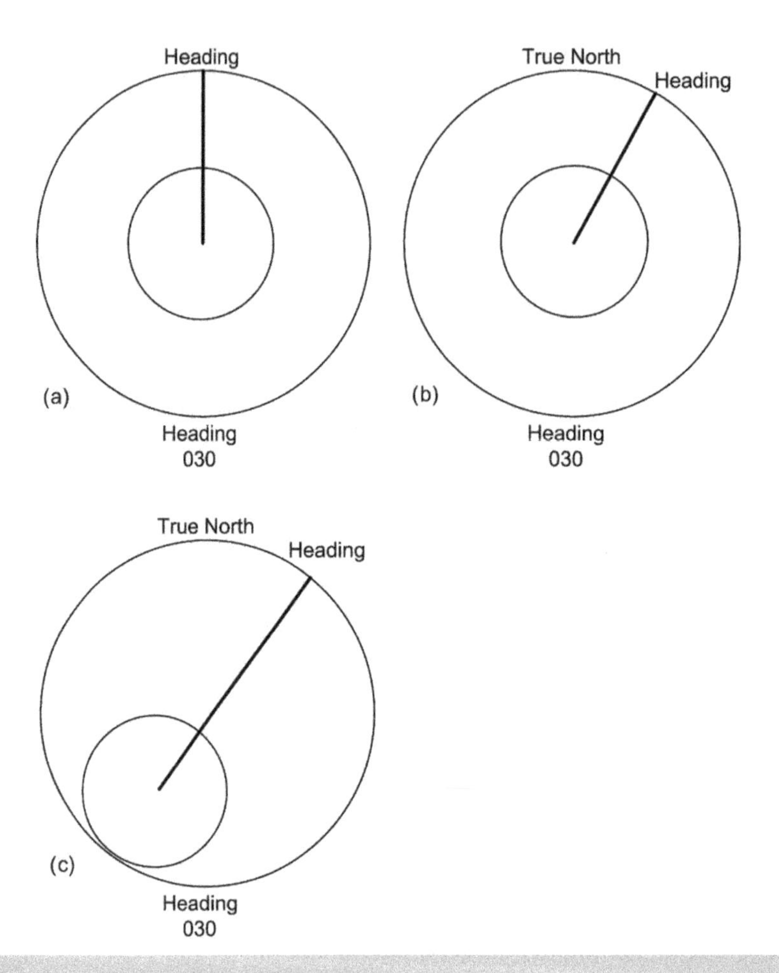

▲ **Figure 2.47** *Radar: True and relative motion*

Automatic Radar Plotting Aid

Automatic radar plotting aid (ARPA) used to be an optional extra for most radar designs but is now included in all radar systems fitted to large vessels. It is a requirement to have ARPA fitted.

Figure 2.48 shows the screen from a Kelvin Hughes MantaDigital Chart Radar. The dashed lines show targets being tracked by the ARPA system. The end of each dashed line represents where the tracked ship will be in 6 min. The tracked target in the top right of the radar display (target 30005) has its data shown in boxes in the far top right of the image.

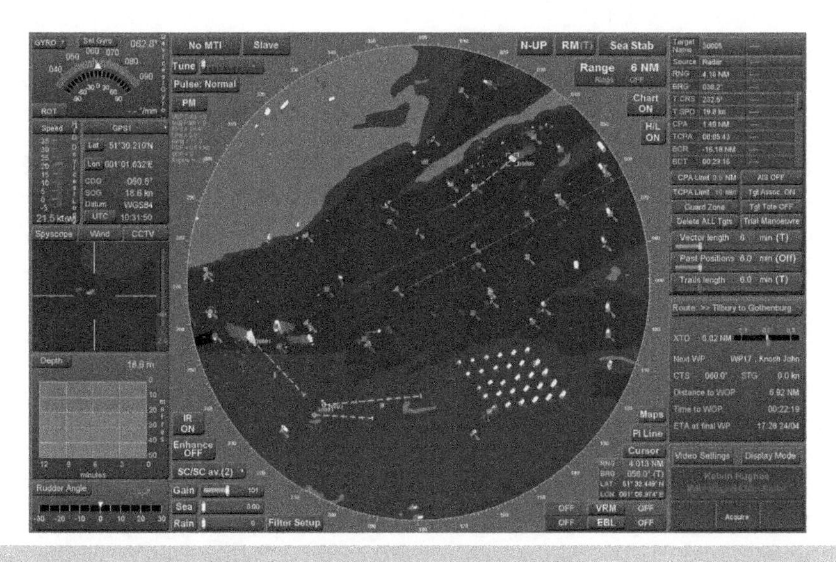

▲ **Figure 2.48** *ARPA: MantaDigital Chart Radar.* © *Kelvin Hughes*

ARPA tracks vessels within range, predicting closest point of approach (CPA), time to closest point of approach (TCPA) and gives the course and speed of other tracked vessels. ARPA is a computer-based tracking system. The echoes received by the radar receiver are digitised after detection. The digital value represents the strength of the echo. The digital value is stored in a bank of RAM. We can convert the x,y format of the RAM into the traditional circular radar display. Figure 2.49 shows this conversion when a target is on a bearing of 090° at 6 miles range.

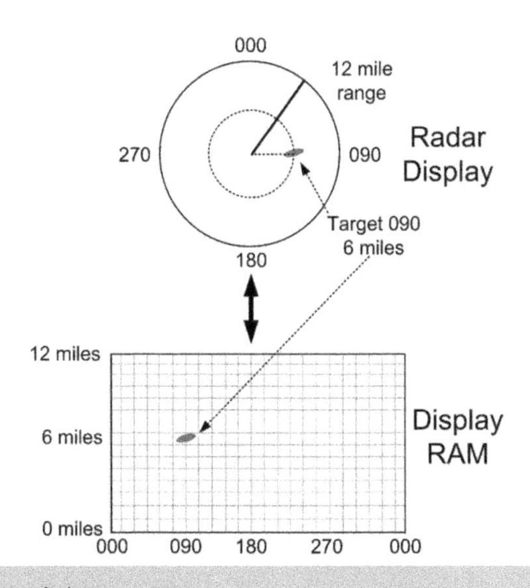

▲ **Figure 2.49** *X, Y to radial conversion*

To help prevent the automatic tracking system from responding to random noise, we use a method called correlation prior to tracking. As an additional benefit, correlation reduces the amount of clutter the operator sees on the radar display. Each cell of RAM represents a small area on the radar display. The whole of the radar display is divided into cells and is assigned a RAM address. As echoes are received, they are digitised and written into the appropriate location in RAM. There are multiple banks of RAM. Each rotation of the antenna uses a different bank of RAM. We will assume there are three banks of RAM. After three rotations each bank is full of echoes. During the fourth rotation the first bank is refilled. We use this rolling system to store the last three rotations in the RAM bank. Clutter and random noise (unwanted echoes) will be stored in each RAM bank in unique locations (Figure 2.50).

If our display RAM or tracking RAM is only updated when an echo is present in three or more correlator banks of RAM in the same location, it is copied into the display/tracking RAM.

This ensures that only 'solid' targets as opposed to random clutter from waves is passed to the display RAM.

Once in the display/tracking RAM, the echo is displayed on the screen. A computer program is constantly running and looking for objects to track. We could define an object suitable for tracking in the display/tracking RAM if it is greater than a certain strength and surrounded by RAM cells that have digital values of near zero. If this object is chosen to be tracked then its X,Y coordinate is stored in the target tracking RAM, with the date and time. After say ten antenna rotations, the target X,Y coordinates with date and time are added to the target tracking RAM. We now have two entries in the tracking RAM for this target, we have two X,Y coordinates with timestamps, we know how big each cell is (varies with selected range) and our speed and course. The

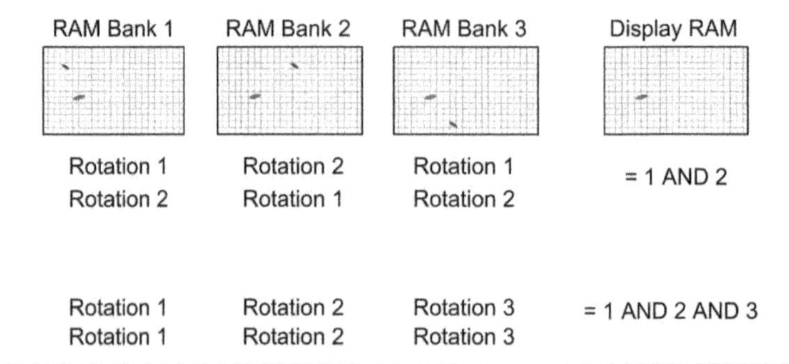

▲ **Figure 2.50** *Correlation with banks of RAM*

computer program calculates the CPA and TCPA from this data. It also paints onto the screen a vector indicating when the tracked vessel will be in 15 min.

Since the computer can calculate cell to cell distances we can set up guard bands and guard rings. If a ship or other object is recorded entering a cell a warning alarm can be given. When at anchor, a fixed point on land can be tracked if we (the vessel) appear to have moved; when we have dragged our anchor an alarm is raised.

Modern radar displays can have AIS and chart data displayed on top of raw, correlated or tracked data. Radar displays can also send the raw video data to other systems that require it such as the VDR or ECDIS.

Figure 2.51 shows radar data merged with AIS and ECDIS data. It shows our ship is steering course 232.5° but our course over ground (COG) is 238.5°.

Data on two targets is shown, VOS Northwind and DSV Seven Pelican. VOS Northwind is from AIS data alone, but DSV Seven Pelican has data from both radar and AIS. The dotted line emanating from each target indicates direction and speed.

Modern radar requires many external inputs to function, each function is automatically checked for validity by the BIT program. If the gyro, log, GPS or AIS data is missing or corrupt, an alarm will be sounded. If the radar control unit detects the scanner has stopped rotating, an alarm will be raised. As with all fault finding, be aware of as many symptoms as possible. Radars are computer controlled devices with BITE that will try to inform you of the failure.

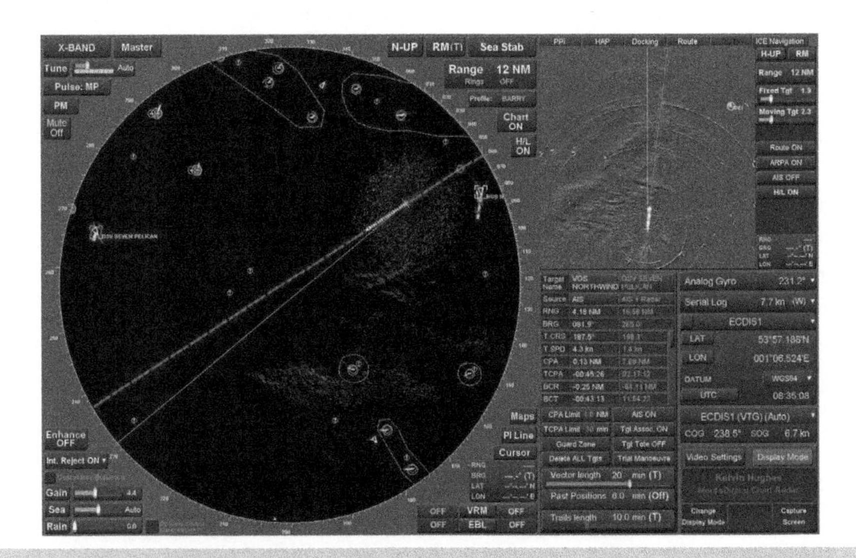

▲ **Figure 2.51** *Merged data: Radar, AIS and ECDIS.* © *Kelvin Hughes*

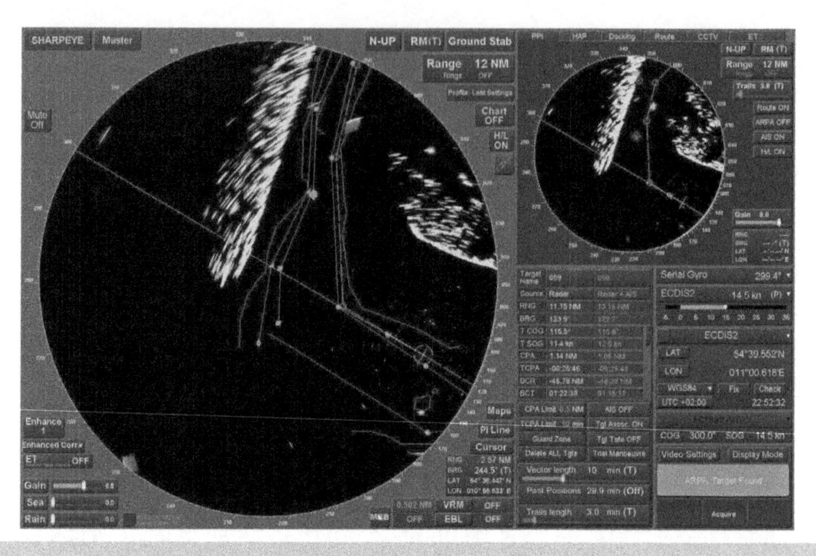

▲ **Figure 2.52** *SharpEye dual PPI display.* © *Kelvin Hughes*

New technology (NT) radar has arrived and allows improved performance and facilities by replacing the magnetron with solid state amplifiers (Figure 2.53). Magnetrons produce an inherently unstable frequency during each pulse due to the high voltage pulse applied to the magnetron disturbing its operation. As the magnetron starts to oscillate, the frequency and phase of its output signal changes, the opposite change occurs at the end of the high voltage pulse when the magnetron is ceasing to oscillate. Solid state amplifiers allow amplification without any frequency variation or phase variation. This gives the opportunity to measure the frequency and phase of the received echoes to determine much more accurately if an echo is clutter or a solid target (solid targets do not change phase or cause an irregular Doppler shift unlike sea and rain clutter). Due to power limitations of solid state amplifiers at 3 and 9 GHz, NT radar power output is limited to a few hundred watts. To overcome this lack of apparent power, longer pulse lengths are used. To increase the strength of the received echo, pulse compression is used. One way of looking at pulse compression would be to imagine the transmitter emitting five 1 μs pulses, 5 μs apart. Any echoes from a valid target would consist of five weak 1 μs pulses, 5 μs apart. If we were to arrange for four parallel delay paths, 5, 10, 15, 20 μs, the outputs of the delays and straight through path summed, the five pulses from a valid target would exit the summing amplifier together, five times stronger. Random received echoes not spaced 5 μs apart would not be added so they would be ignored.

Due to minimum distance requirements for target detection from IMO, NT radar requires a very sophisticated pulse compression scheme when compared to the simple one described here. This allows the NT radar to detect close targets and distant targets with equal performance (Figure 2.53).

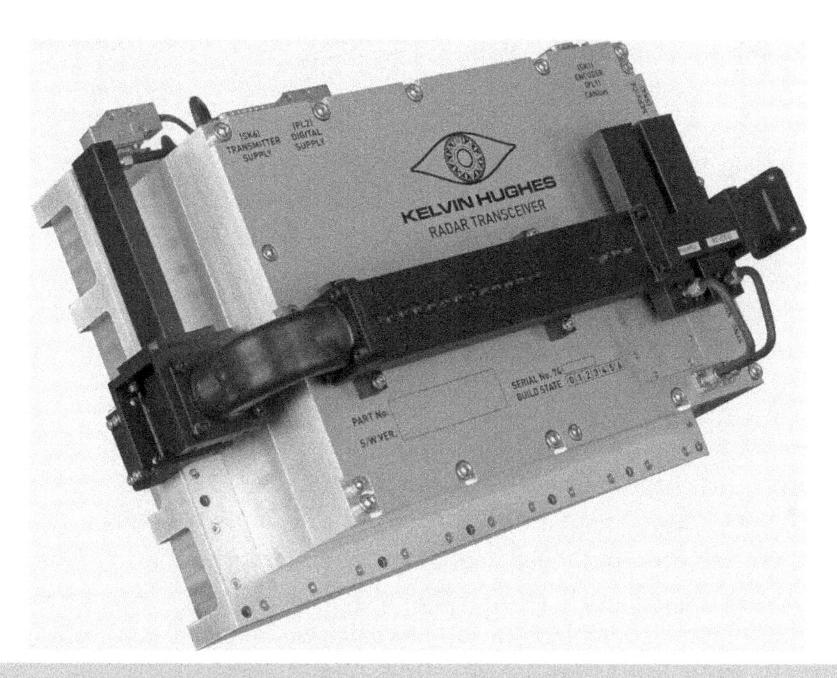

▲ **Figure 2.53** *SharpEye X-band solid state transceiver.* © *Kelvin Hughes*

Electronic Chart Display and Information System

Electronic Chart Display and Information System (ECDIS) has revolutionised marine navigation. It is an electronic chart display system built using a series of approved components. Charts can be displayed in one of two modes: raster or vector. The Raster Chart Display System (RCDS) uses charts that are scanned versions of traditional paper charts. The user can zoom in and out but the level and amount of information available to the user remains the same. A limited amount of automation is available using RCDS.

Vector charts give full ECDIS capability. Vector charts store in digital format all of the data required to display an Electronic Navigation Chart (ENC). An ENC is a formally released digital data set, compliant to international standards and of sufficient quality to be used for navigation. The digital format of the ENC allows the ECDIS operator to zoom in and out to see more or less detail. Areas and items of interest can be clicked on to show greater detail. The digital data is supplied in layers. Each layer contains different classes of information of varying importance to the navigator. The ECDIS system allows different layers of information to be available to the navigator, allowing the system to be configured to display only relevant data.

Figure 2.54 shows a typical ECDIS display. The ECDIS system is a PC based chart display system. It has inputs from GPS, gyro and log. ECDIS displays the ship's current position as determined by GPS. Using the speed and heading input, ECDIS shows a line of predicted track, enabling the navigator to judge future positions. ECDIS contains voyage planning software to allow the navigator to input start and end points, complete with way points as necessary. A program can be run which checks that the intended route is safe for the vessel: depth and distance from hazards. Modern ECDIS can output course setting commands to an autopilot, raising the possibility of a vessel completing a voyage with no human course correction commands. Although ECDIS are simple display devices, they have a fundamental role in the safe navigation of a modern vessel.

To increase the availability of the ECDIS functionality on board, most ECDIS equipped vessels have a dual ECDIS installation, each ECDIS powered by individual uninterruptible power supplies (UPS). To enable easy chart updates, the ECDIS systems will be networked together and connected to the communications system to enable chart corrections to be downloaded from an official ENC agent. Radar data can be superimposed onto the ECDIS display, along with AIS data, potentially giving a complete but complex display. See MFD.

An increasing trend is to have general purpose workstations that can be switched to display the following information: radar, ARPA, AIS and ECDIS in any combination to suit the officer of the watch. All of this data is now recorded on the VDR. Figure 2.55 shows ECDIS integration with bridge equipment.

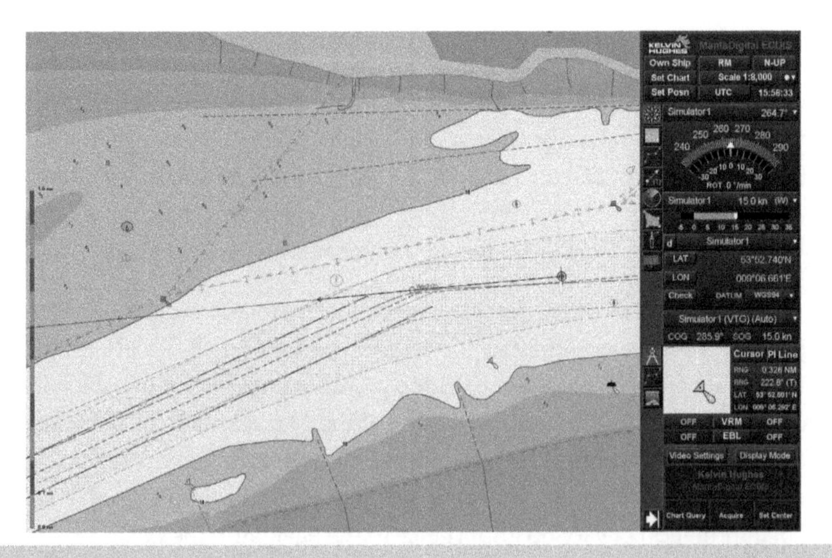

▲ **Figure 2.54** *Typical display MantaDigital ECDIS. © Kelvin Hughes*

▲ **Figure 2.55** *Integrated bridge: SharpEye Radar plus ECDIS. Reproduced by kind permission of Austal*

As ECDIS is now an integral part of the control of the vessel, fault finding has become very important. ECDIS requires data from many other navigational systems, each data feed is checked by the ECDIS to ensure it is present and error free. When a data feed fails, the ECDIS system will raise an alarm; the navigator could use the alternate ECDIS if necessary. If the ECDIS failure was due to a common input, that is, the gyro feeding both ECDIS, then the ECDIS will require switching into dead reckoning (DR) mode, until the fault is cleared or the standby gyro enabled. As most data links between navigational equipment are NMEA 0183, we would follow the wiring from the data source (gyro, AIS, GPS, etc.) to the ECDIS, measuring the ECDIS input data to find the fault.

Voyage Data Recorder

Due to a number of serious accidents at sea, most large vessels are now mandated to carry a voyage data recorder (VDR) to record significant commands and ship performance data. Commands from the bridge to engine/engine room are recorded along with helm, speed, GPS coordinates, AIS and radar data, and bridge conversations via microphones.

When the VDR ruling was implemented it was found that many older ships had analogue communications paths. It would have been very expensive to install a VDR into an older vessel due to the cost of converting so many analogue signals into digital. To overcome this problem, a simplified VDR (SVDR) was allowed to be installed on older vessels where only a subset of the full data set is recorded. Figure 2.56 shows the main data gathering and recording unit from Kelvin Hughes.

VDR and SVDR are just computer controlled recording devices that record the activity listed above. The storage system can be mounted in such a way as to float free when the vessel sinks. Some VDRs allow a removable disk to be hand-carried when abandoning a vessel.

Figure 2.57 shows the crash survivable module (CSM) from Kelvin Hughes. It can be installed to remain fixed to the ship or to float free. It contains a copy of all the data and an audio beacon to help find it when submerged.

Both VDR and SVDR verify their inputs by running BIT and can raise an alarm if a required input becomes corrupt or open circuit.

A technician authorised by the VDR/SVDR manufacturer verifies that the unit is fully working and correctly maintained each year.

Figure 2.58 shows the typical configuration of a VDR. Fault finding on VDRs requires us to take note of warning messages regarding missing inputs, tracing back to the origin of the signal as necessary. VDRs are surveyed annually.

▲ **Figure 2.56** *VDR main unit. © Kelvin Hughes*

▲ **Figure 2.57** *VDR survivable unit. © Kelvin Hughes*

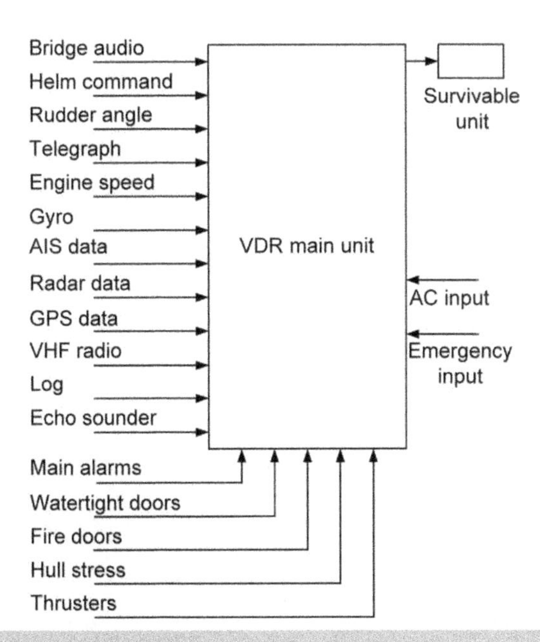

▲ **Figure 2.58** *VDR system overview (many inputs recorded)*

Navtex

Navtex is a MF radio broadcast system on 518 kHz used to transmit navigational, meteorological and safety messages to vessels. It uses narrow band direct printing (NBDP) to print or display important messages for the seafarer. The transmissions all share a common frequency of 518 kHz. To prevent interference from other Navtex stations, messages are sent during 10-min slots every 4 h. Each coast radio station is allocated a 10-min slot by its national administration, which is agreed on at an international level. The range of MF Navtex signals is up to 400 miles.

The format of Navtex messages includes a message type indicator:

A: Navigational warnings,

B: Meteorological warning,

C: Ice reports,

D: Search and rescue information and pirate attack warnings,

E: – Z.

Navtex receivers can be set to ignore certain types of message, if the Master deems them not relevant, but types A, B and D messages cannot be ignored.

Each Navtex message contains a two digit serial number from 01 to 99. This allows a message received intact to be displayed only once. Messages with a serial number 00 will always be printed. All international Navtex messages should be transmitted in English; the frequency 490 kHz is available for national language transmissions.

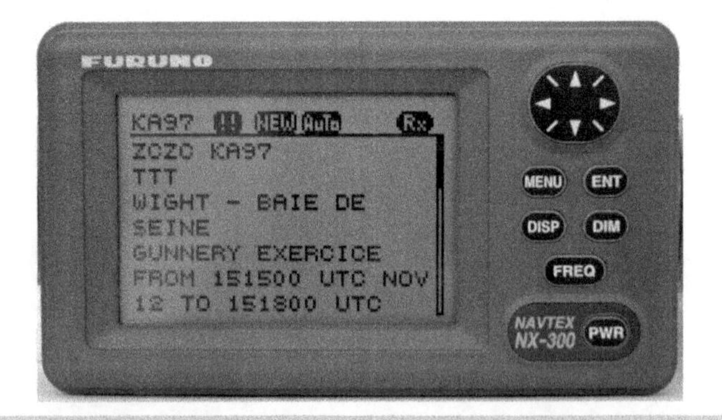

▲ **Figure 2.59** *Furuno Navtex receiver*

Navtex receivers are simple, direct printing devices. They contain a microprocessor with the ability to diagnose most faults during the power up sequence. The Navtex receiver antenna can be a short whip or a long wire. Ensure that the antenna connection to the receiver is in good condition and any antenna insulators are free from salt and dirt. Figure 2.59 shows a typical Navtex receiver with LCD screen.

Fault Finding in Bridge Equipment Systems

Fault finding navigational aids on board a modern merchant vessel, with its collection of advanced equipment, has become a systems level task. Individual items of equipment, which contain computer controlled and surface-mount printed circuit board designs are difficult to repair ashore let alone at sea. With the difficulty of repairing equipment down to component level we must be satisfied with system level repairs. This requires us to understand the interconnections of all the electronic equipment on board. We need to understand how each item of equipment communicates with others. Figure 2.60 shows a partial set of bridge equipment with interconnections. The complexity of interconnections in Figure 2.60 indicates to us that we need a very good understanding of the source and destination of each data line.

We need to access the system diagrams supplied with the vessel or create our own diagrams to enable us to understand the system level connectivity. When equipment raises an alarm, it will either flag up an internal failure or a signal input failure. If an input signal has failed we need to check if other devices indicate the same signal being faulty. If multiple equipment flags the same failing input, it is reasonable to assume the source

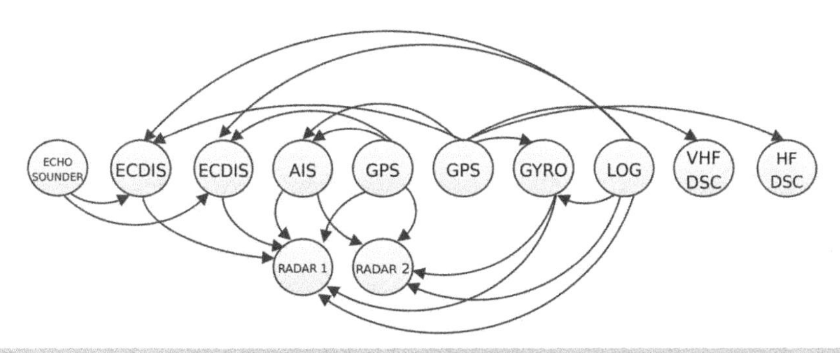

▲ **Figure 2.60** *System fault finding (sample interdependency)*

of the signal is faulty. We then need to check the source of the signal. For example, if the radar reports loss of gyro, if no other equipment reports a failure then the fault could be (a) the radar gyro input, (b) gyro output to radar, (c) cable in between.

If however two or more items of equipment report the loss of the gyro signal, then the gyro could be the main suspect.

3

RADIO COMMUNICATIONS

Radio communication, the transmission and reception of signals over a distance, is of vital importance to vessels at sea. Vessels need to communicate with each other and shore stations for safety of life at sea (SOLAS) purposes. The SOLAS convention dictates what communication equipment and methods are required. Before we discuss transmitters and receivers, we need to understand what we are transmitting and how radio waves travel.

Radiation and Propagation

When an alternating current (AC) travels along a wire, electromagnetic (EM) fields emanate from the wire. These EM fields travel or radiate from the wire. When the EM field reaches another conductor, a voltage is induced into this second conductor. Experiments by Hertz showed that simple wire antennas would transmit and receive EM waves.

It can be difficult to 'launch' EM waves or radio waves from a conductor. It is most efficient if the length of the antenna is related to the wavelength of the AC signal. If two wires, end to end, are energised with AC, with a frequency such that the total length of the antenna is half the wavelength of the applied AC frequency, the wire will radiate.

Figure 3.1 shows a half-wave dipole for an AC frequency of 500 kHz. This formula $\lambda = c/f$ gives us the wavelength in metres for a frequency in Hertz where c is the speed of light in metres per second. For a frequency of 500 kHz the wavelength is 600 m, so a half-wave dipole would be 300 m in length.

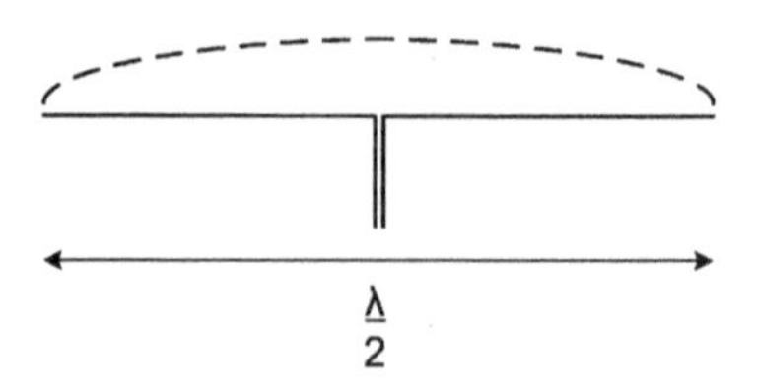

▲ **Figure 3.1** *Half wave dipole*

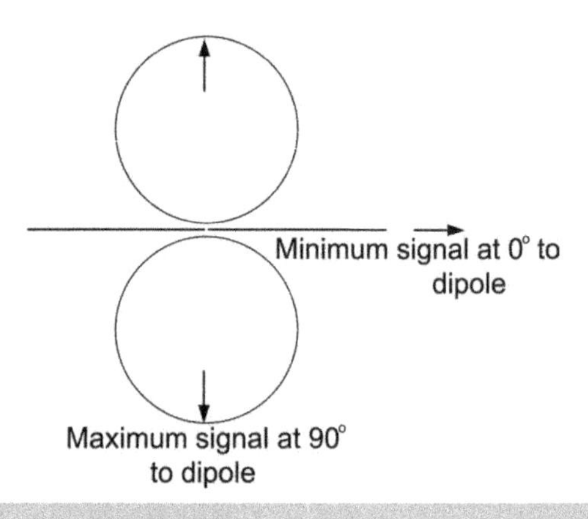

▲ **Figure 3.2** *Radio waves emanating from the dipole*

Figure 3.2 shows the radio waves emanating from the dipole. If the frequency and the length of the antenna matched, the current measurable along the ½ wave dipole would be as shown by the dotted line in Figure 3.1.

The maximum current would be in the centre of the antenna, minimum at the two ends. Ohm's law tells us that maximum current occurs at minimum resistance, so the centre of the dipole exhibits minimum resistance. Ideally the dipole will be designed to have an impedance of 50 Ω. This would enable us to connect our transmitter to the antenna via a 50 Ω impedance cable or transmission line. As described in the principles section, maximum power is transferred from a source to a load if both have matching impedances. A variance between the transmitter, transmission line or antenna gives rise to a mismatch. This mismatch can be serious at radio frequencies because radio waves reflect back from a mismatch. The reflection reduces the energy emitted from the antenna, and the reflection is passed back to the transmitter output stage causing overheating. This requires us to consider antenna tuning. To tune an antenna to match the impedance to the transmission line means altering the electrical length of the antenna by adding C or L as necessary to reduce the mismatch and reflection. Antenna tuners enable an antenna to be given more or less L or C to cause a match. If the transmitter frequency was changed, the length of the antenna would no longer

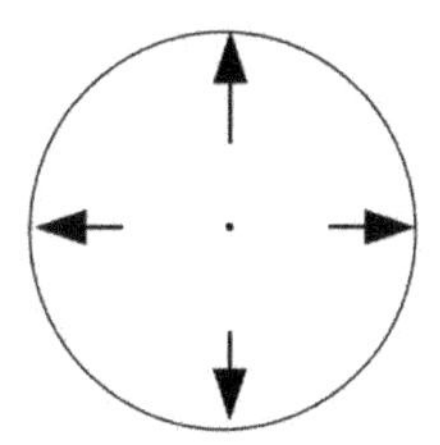

▲ **Figure 3.3** *Vertical dipole has omnidirectional radiation*

be matched, causing loss of emitted power. The antenna tuner would require retuning after each frequency change. A broadband antenna removes much of this difficulty by keeping match over a wide range of frequencies.

On many occasions a full size half-wave dipole is not practical and alternative designs must be considered. Also, the omni-directional properties of a half-wave dipole are insufficient. A popular antenna is the quarter-wave whip. Effectively this is half of a half-wave dipole. Figure 3.4 shows a quarter-whip antenna. These are popular with handheld equipment. Figure 3.3 shows a vertical antenna with equal signal strength over 360 degrees.

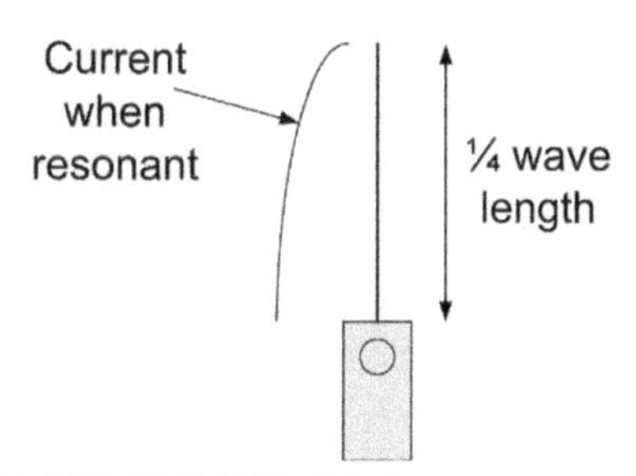

▲ **Figure 3.4** *Quarter wave whip antenna*

To obtain a directional antenna, reflective properties of adjacent conductors are used.

In Figure 3.5, we see a half-wave vertical dipole driven by energy from a transmitter. Within quarter of a wavelength distance from the half dipole we have an un-driven half dipole which receives radio wave energy from the driven dipole, the induced current in the reflector causes radio waves to be emitted from the reflector dipole.

Down range from the driven dipole, the radio waves merge into one. The merged signal will be more powerful in one direction and less powerful in the opposite direction. The addition of directing elements on the opposite side to the reflector increases the strength of signal further and makes the antenna more directional.

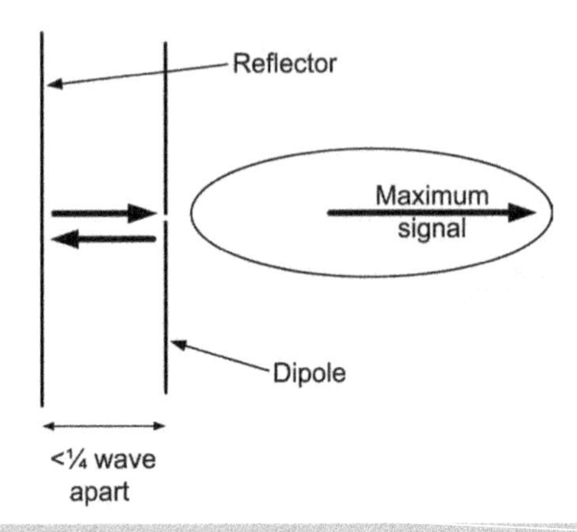

▲ **Figure 3.5** *Directional antenna*

Antennas are reciprocal in operation, that is, an effective transmitting antenna will be effective for reception.

Radio waves propagate (travel) from the transmitter antenna to the receiver antenna in a manner dependent upon the frequency being transmitted. The radio spectrum can be conveniently divided into bands which have unique characteristics. Table 3.1 shows the radio spectrum from 3 kHz to 300 GHz.

Radio waves can be considered to propagate in three different modes: (1) ground wave, (2) sky wave and (3) direct line of sight.

The lower frequency bands (VLF and LF) all use ground wave; medium frequency bands (MF and HF) use a mix of ground wave and sky wave dependent upon time of day, season and sun spot cycle. Higher bands (VHF and above) use direct line of sight propagation. All frequency bands can be affected by anomalies in sun activity.

Table 3.1 *The radio spectrum, 3KHz–300GHz*

VLF	Very low frequency	3–30 kHz
LF	Low frequency	30–300 kHz
MF	Medium frequency	300–3000 kHz
HF	High frequency	3–30 MHz
VHF	Very high frequency	30–300 MHz
UHF	Ultra high frequency	300–3000 MHz
SHF	Super high frequency	3–30 GHz
EHF	Extremely high frequency	30–300 GHz

Ground wave propagation is where the transmitted wave clings to the earth's surface as it travels around the globe. This gives a very stable radio signal but the long wavelength requires very large antennas (many kilometres long) and very high transmitter power (megawatts). The distance these waves travel is dependent upon transmitter power.

Sky wave propagation is where the transmitted wave is reflected by bands of ionised gas in the upper atmosphere. Surrounding the earth are distinct bands of ionised gas which reflect or absorb radio waves (Figure 3.6). These layers named D, E and F vary in height above the earth's surface controlled mainly by the activity of the sun. The sun emits solar radiation which varies in strength over a 11-year cycle (sun spot cycle). The sun's energy increases or decreases the effect of the layers. At night time, the lack of sun light causes the D layer to disappear, reappearing at dawn. The D layer absorbs much medium frequency transmissions during the day time; at night time, with the D layer removed, many distant MF radio stations can be heard. During the day time, medium frequencies benefit for ground wave propagation, at night time, sky wave propagation gives much greater range. High frequencies have little ground wave propagation and rely heavily on sky wave propagation.

At VHF frequencies and above, direct line of sight is the main mode of propagation. Ground wave propagation is so small as to be ineffective. Sky wave effects are virtually zero. VHF and above tend to pass straight through the D, E and F layers.

Direct line of sight implies that the receiver antenna needs to be 'seen' by the transmitter antenna. This is certainly true for UHF and above, but for VHF, the radio waves do

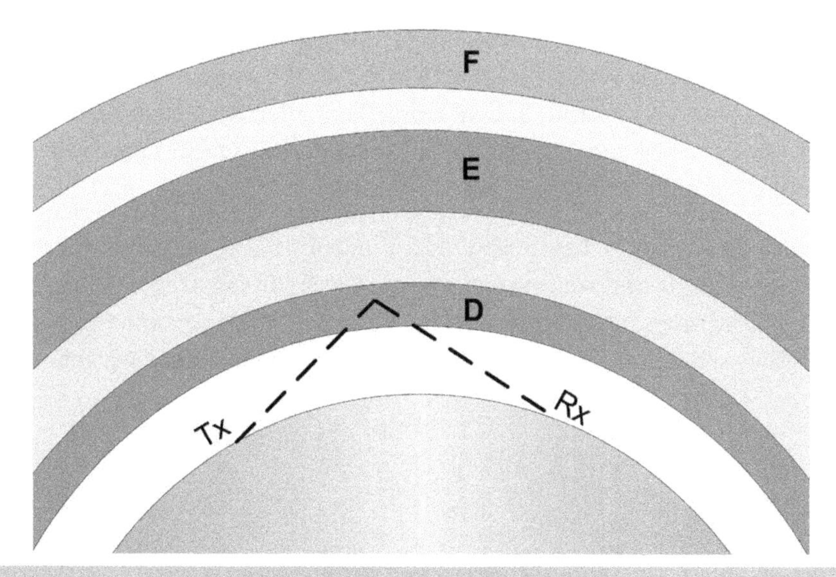

▲ **Figure 3.6** *Layers in ionosphere: D, E & F*

partially follow the curve of the earth surface for a short distance, allowing VHF radio communication slightly further than the visible horizon. UHF and above are line of site. This is easily demonstrated with the GPS navigation system where the satellites have to be 'seen' above the observer's horizon before the receiver will receive any signals.

As can be seen from figure 3.6, the ionosphere is complex, changes dramatically during the day and night and from season to season. To ensure vessels can communicate when out of sight of land, a range of frequencies or wavebands are used. These frequencies work best at different times of day and season. The High Frequency (HF) bands are: 2 MHz, 4 MHz, 6 MHz, 8 MHz, 12 MHz and 16 MHz.

Very High Frequency (VHF) band is 156 MHz. Precise frequencies are specified for ship calling or shore station calling. If you instigate a distress call on a GMDSS fitted vessel, distress messages will be sent out on each frequency band in turn. It is to be noted that Inmarsat GMDSS via satellite is not guaranteed above about 80 degrees north or below 80 degrees south of the equator. Iridium GMDSS is truly global in coverage.

Amplitude and Angle Modulation

We now have the understanding of how a transmitting antenna radiates EM waves to a receiving antenna. We now need to communicate a message using these radio waves. The simplest method is to turn the radio waves on and off in an agreed pattern. The most recognisable method using this technique is Morse code. Using this code, a transmitting antenna would be fed with bursts of radio frequency (RF) generating radio waves, which are picked up by the receiving antenna and can be decoded by a human operator. The block diagram of a transmitter capable of transmitting Morse code is shown in Figure 3.7.

An RF oscillator generates the required radio frequency, the power amplifier boosts the level of signal to the desired level and a Morse key is connected to control when the power amplifier sends the high power signal to the transmitting antenna. The receiving mechanism is shown in Figure 3.8, where the receiving antenna is connected to a tuned

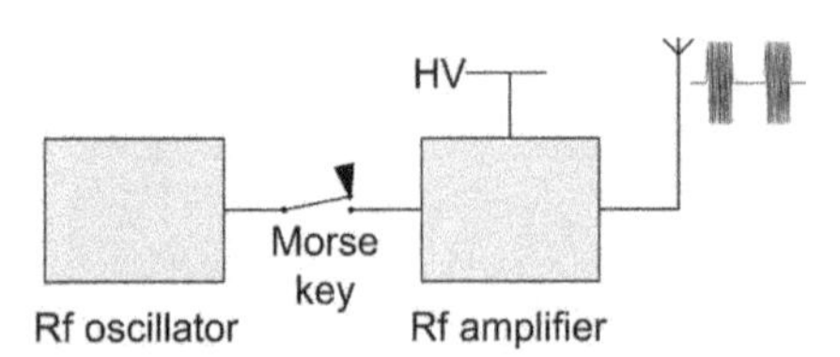

▲ **Figure 3.7** *Transmitter for morse code*

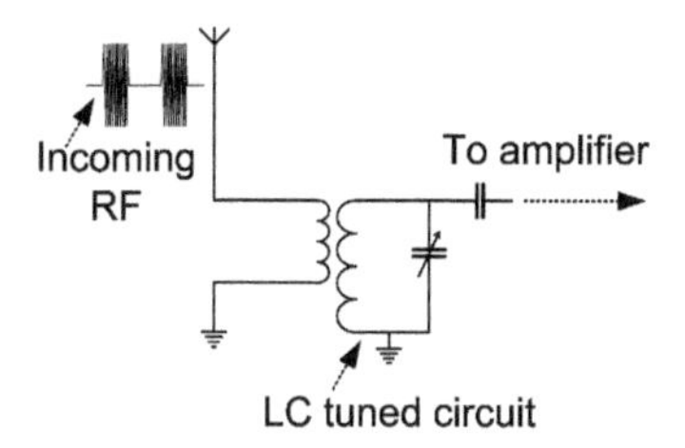

▲ **Figure 3.8** *Receiver front end tuning*

circuit consisting of an inductor and capacitor. We know that an LC circuit will oscillate at a frequency of

$$f_o = \frac{1}{2\pi\sqrt{LC}}\text{Hz}$$

when stimulated with an AC voltage of that frequency. The signal induced into the receiving antenna is coupled to the LC circuit and passed to an amplifier. The amplifier increases the very small received signal into a useful level. How the signal is handled further depends on the nature of the modulation system chosen.

We need to impress the message onto the signal (voice or data) to enable the radio wave to 'carry' the message to the receiver. Adding a message to the transmitted wave is called modulation.

Modulation changes the continuous radio frequency wave in a way that the receiver can 'detect' the message easily. A number of modulation schemes have been developed to modulate the transmitted wave.

- Amplitude Modulation (AM): Single Side Band (SSB).
- Angular Modulation: Phase Modulation (PM), Frequency Modulation (FM).

AM allows for voice signals to modulate the carrier wave. This is shown in Figure 3.9. The audio frequency is used to change the amplitude (size) of the transmitted signal.

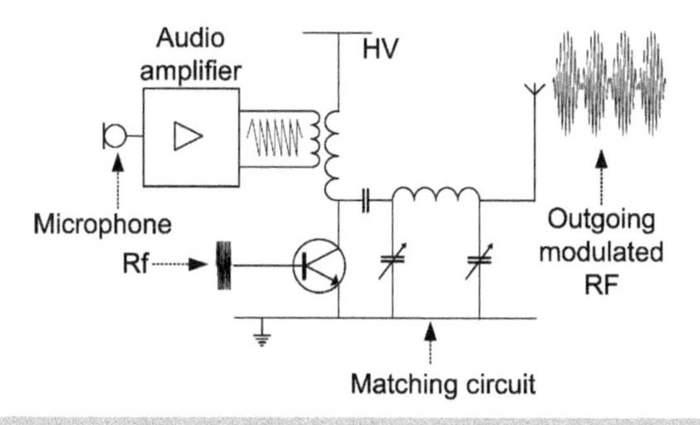

▲ **Figure 3.9** *Amplitude modulated transmitter*

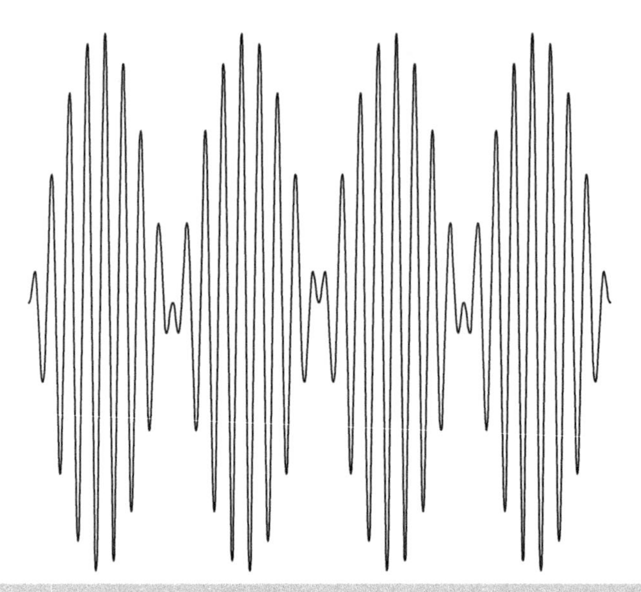

▲ **Figure 3.10** *Amplitude modulation (by time)*

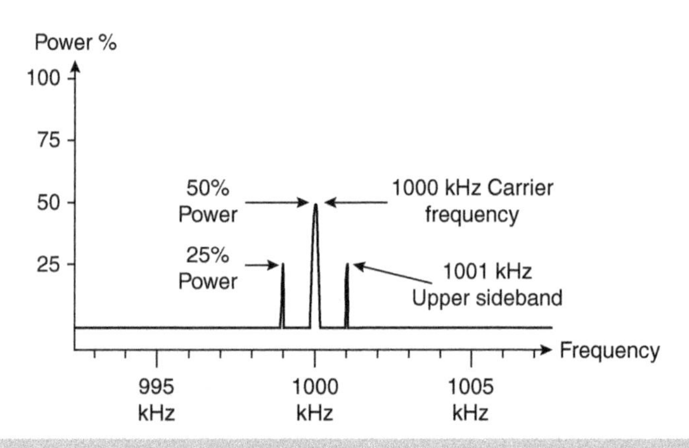

▲ **Figure 3.11** *Spectrum analyser view of amplitude modulation*

This allows the carrier wave to 'carry' an impression of the audio frequency 'embedded' onto the carrier wave. The section on receivers shows how we extract or detect the embedded audio from an AM signal.

Figure 3.10 shows the effect of an audio waveform modulating an RF waveform. The shape of the AF waveform can be 'seen' in the variation in amplitude of the RF wave.

Figure 3.11 shows a carrier wave modulated by a continuous tone of 1 kHz. Two side bands can be seen, a lower one at 999 kHz and upper side band at 1001 kHz.

Figures 3.10 and 3.11 both show the transmitted signal from different viewpoints, Figure 3.10 shows the signal travelling from left to right and Figure 3.11 shows the signal 'coming' towards us. The receiver enjoys the combined signal and detects or demodulates it to extract intelligence from it.

The combined result the receiver listens to is the AM signal as seen in Figure 3.10. However, the message is only carried in the side bands. The carrier contains 50% of the transmitted power and each side band contains 25% of the transmitted power. We could arrange for a much more efficient radio communications system if we could remove the carrier and just used the side bands. Each side band is a mirror image of the other, so we could remove one side band and use the transmitter power amplifier to use all of its power to amplify one side band. This gives SSB communication links greater range and efficiency.

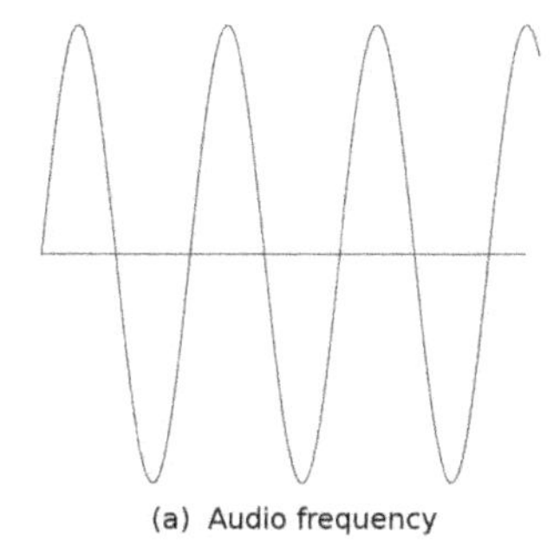

(a) Audio frequency

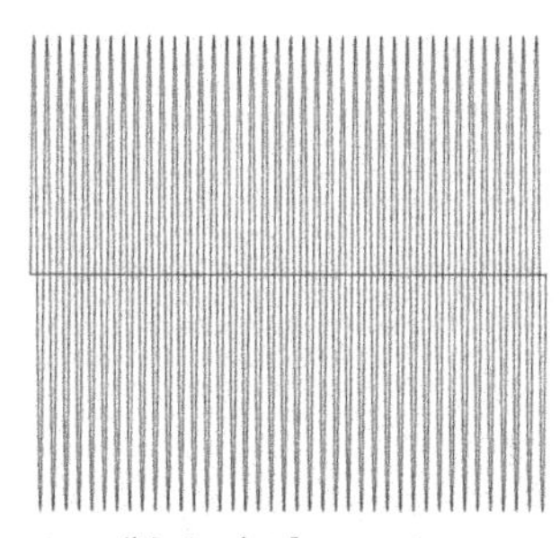

(b) Carrier frequency

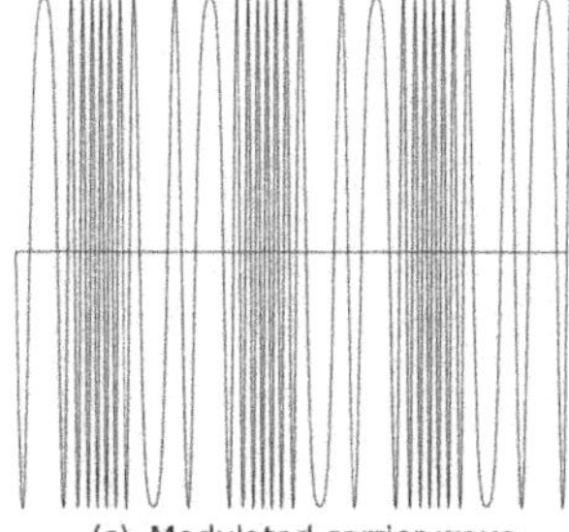

(c) Modulated carrier wave

▲ **Figure 3.12** *FM modulation waveforms*

A further benefit of SSB is spectrum utilisation. An SSB signal takes up lesser spectrum than double side band plus carrier. Most long range voice communication on HF and MF uses SSB.

If we kept the carrier wave at constant amplitude but varied its frequency in response to a voice signal, we would have angular or frequency modulation.

Frequency and phase modulation are mathematically related. For voice communications FM is used in the majority of cases. Figure 3.12 shows FM transmitter waveforms. The modulating signal f_m of 1 kHz is modulating a carrier frequency f_c of 1 MHz. The transmitted signal frequency can be seen to change in sympathy with the message signal. The spectrum analyser view of the FM transmitted wave is shown in Figure 3.13. The side bands can be seen at 1 kHz increments above and below the carrier frequency. The number and height of side bands is determined by the modulation index. There are an infinite number of side bands in FM, but the power in these side bands reduces rapidly. Carson's rule tells us that the effective bandwidth of an FM transmission is $2(D_{max} + M_{max})$Hz, where D_{max} is the maximum frequency of deviation and M_{max} is the maximum modulating frequency.

For example, if the maximum frequency deviation is ±5 kHz and the maximum audio frequency is 3 kHz then: bandwidth 2(5+3) = 16 kHz (approximately). International agreements dictate that marine VHF transceivers use ±5 kHz maximum deviation and a channel spacing of 25 or 12.5 kHz. Due to sensitivity to noise, the VHF FM transmitter audio signal is emphasised at a rate of 6 dB/octave, which means the signal amplitude is doubled every time the frequency is doubled. The process is reversed in the receiver. The effect of this pre- and de-emphasis is to give the FM system a very low level of noise when listening to stations. We can improve the noise performance by increasing the amount of frequency deviation, but this would increase the bandwidth significantly.

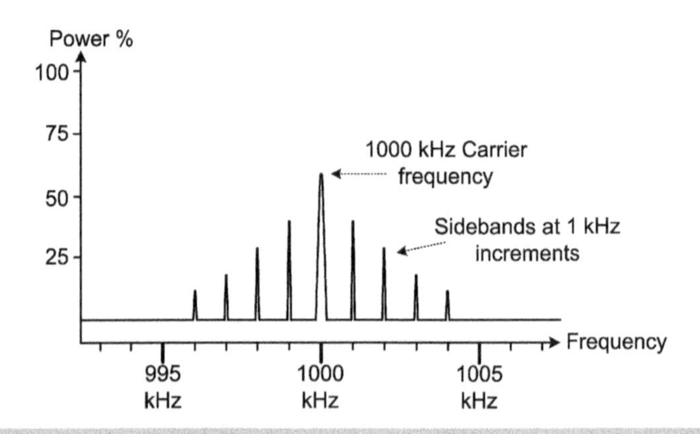

▲ **Figure 3.13** *FM modulation sidebands*

This is used in FM broadcasting between 88 and 108 MHz where maximum frequency limit is 15 kHz and the deviation limit is ±75 kHz. This creates a very good broadcast signal but Carson's rule informs us that the bandwidth is now 180 kHz which just fits into the 200 kHz broadcasting channel allocation.

Radio Transmitters

To generate the signals needed to drive electromagnetic waves out from an antenna, careful consideration is given to the circuits used. For effective radio transmission we require accurate and stable frequencies. Traditionally these stable and accurate frequencies would be created with the aid of a quartz crystal. When these crystals vibrate they generate very small AC voltages. The frequency of vibration is determined by the physical size of the crystal. By keeping the crystal at a constant temperature, its size would be constant and the frequency constant. Newer transmitters no longer use one crystal for each transmitted frequency but instead use phase locked loop (PLL) and a digital integrated circuit divider. By changing the digital divide ratio any frequency within a range can be selected.

Figure 3.14 shows the concept. This arrangement is so common now that complete frequency synthesisers can be obtained on one chip. The small RF signal needs to be amplified and connected to the antenna. How we amplify the signal depends upon how we are going to modulate it. Different methods are used for each type of modulation scheme.

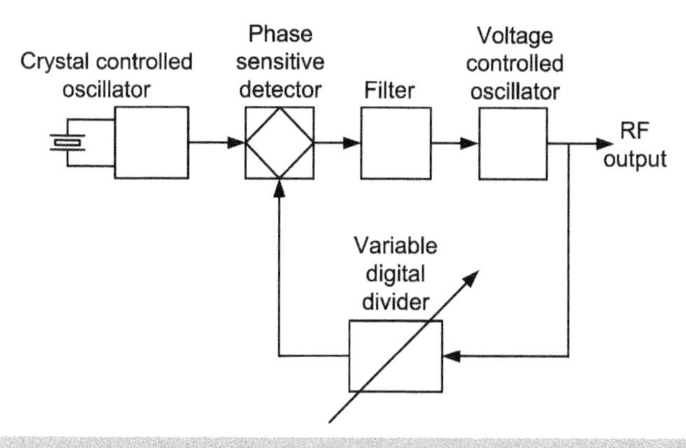

▲ **Figure 3.14** *Phase locked loop synthesiser*

Transmitting amplitude modulation

As previously discussed, Figure 3.9 shows a very simple AM transmitter where the voice modulating signal is used to increase or decrease the voltage connected to the final amplifier immediately prior to the antenna. We can see in Figure 3.10 the effect of AM on a carrier f_c.

Transmitting single side band

Single side band as discussed above eliminates the carrier and one single band. Typically the lower side band is removed allowing the upper side band to be transmitted. As the upper side band now wholly carries all of the intelligence of the modulating signal, it must be amplified linearly without distortion.

Figure 3.15 shows the oscillator signal being fed into a balanced mixer where the carrier is filtered out. Next, the lower side band is removed by a filter. Figure 3.16 shows a typical filter.

Figure 3.17 shows the SSB signal being changed to a higher frequency ready for amplification by a linear amplifier.

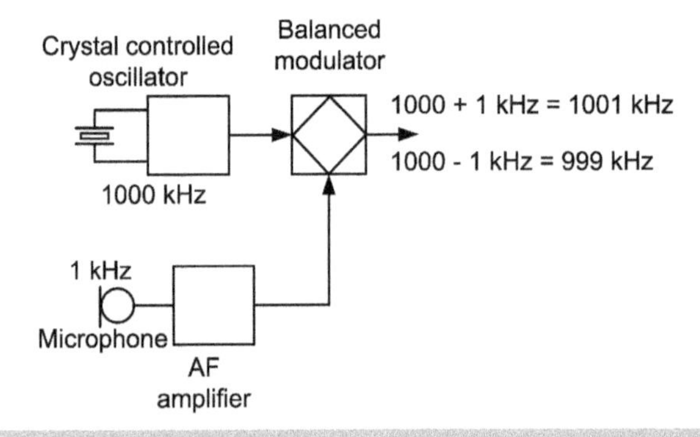

▲ **Figure 3.15** *Balanced mixer/modulator*

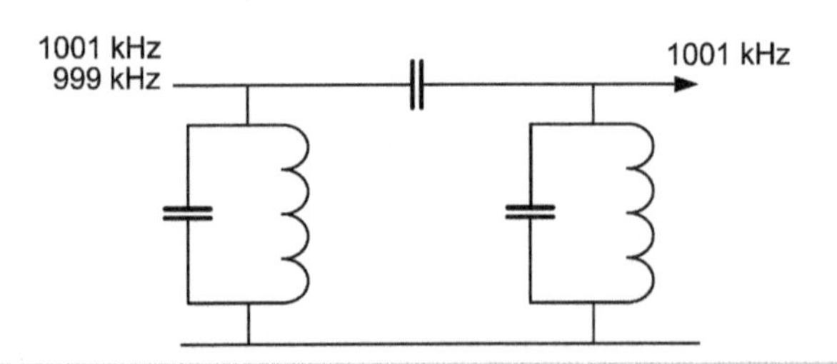

▲ **Figure 3.16** *Upper sideband filter*

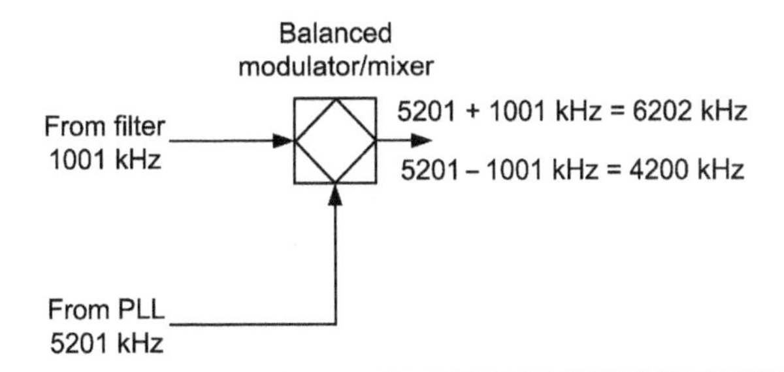

▲ **Figure 3.17** *Frequency convertor*

All this has taken place at a low frequency to make it easy to filter out the side band. We now need to mix this single side band signal with a higher frequency to reach the final transmitted frequency. This is done with a series of frequency changes. Finally the upper side band signal is fed into the linear amplifier to reach final transmit strength and connected to the antenna via a matching unit.

Transmitting frequency modulation

An FM transmitter is simple. A crystal oscillator has a Varactor connected across its crystal. A Varactor is a variable capacitance diode, reverse biased, its capacitance varies with the size of the reverse voltage. This small change in capacitance connected to the crystal is sufficient to 'pull' the frequency of the crystal.

The microphone is connected to an audio amplifier with audio limiters. We need the limiter to prevent someone with a loud voice causing the transmitter to exceed the frequency deviation and interfere with adjacent channels. After the audio amplifier is the 6 dB/octave pre-emphasis circuit that is used to reduce noise in the received signal. The Varactor or varicap diode is modulated with this audio signal. The FM signal is now multiplied in frequency, amplified and connected to the antenna via the usual method. This can be seen in Figure 3.18.

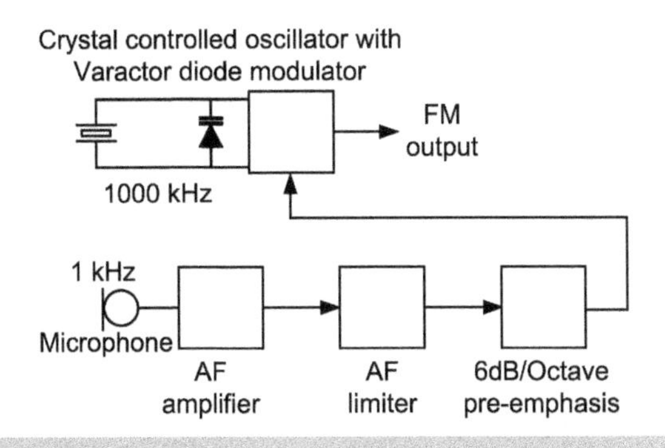

▲ **Figure 3.18** *FM transmitter*

Radio Receivers

To tune into and decode the intelligence being transmitted requires different techniques to suit each style of transmission.

Receiving amplitude modulation

Figure 3.19 shows an AM receiver. It consists of a tuned circuit which will oscillate at a frequency determined only by the value of the *L* and *C* using the formula:

$$f_o = \frac{1}{2\pi\sqrt{LC}} \text{Hz}$$

This allows the radio receiver to select the signal of interest. This selected signal is amplified in the RF amplifier and connected to a diode; the detector diode is used to rectify the radio signal converting it into direct current (DC). The voltage output depends upon the transmitted modulated signal. If we remember, with AM, the transmitted/received signal is changing in size (modulated) due to the audio signal. This means that if we rectify the received AC signal, we recover the audio signal. This used to be called 'detecting' the signal using a diode detector.

As frequencies get higher, it becomes difficult to separate the many transmitted signals with such a simple receiver, whose selectivity relies upon a simple LC circuit. A better approach would be to use the superheterodyne principle. Figure 3.20 shows a typical superheterodyne receiver. This configuration allows for the down shifting of the RF to a lower frequency that can be more easily amplified using circuits that are more selective. At 465 kHz, the intermediate frequency (IF) amplifier can use a succession of

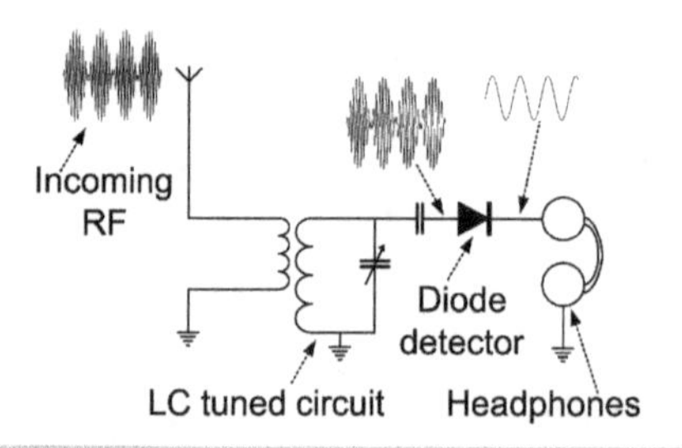

▲ **Figure 3.19** *AM Receiver with diode detector*

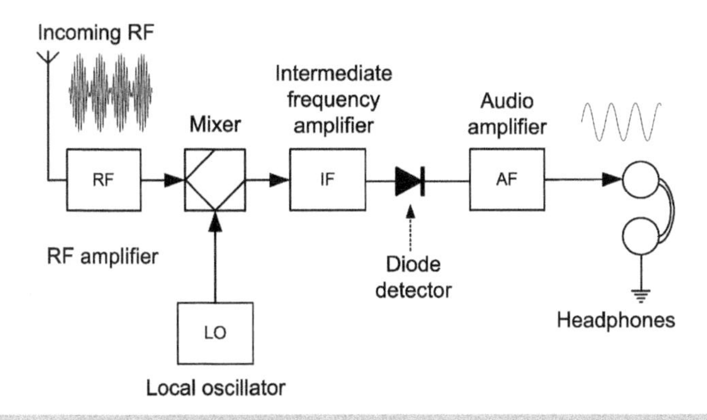

▲ **Figure 3.20** *AM Superheterodyne receiver*

simple LC circuits to gain sufficient selectivity to allow the required frequency through without adjacent channels interfering. To obtain a constant IF frequency, we use a local oscillator (LO) and mixer. The incoming desired RF signal enters the mixer along with the LO signal, which is exactly 465 kHz higher in frequency. These two frequencies mix together, giving sum and difference products. A sum product would be 1000 kHz plus 1465 kHz creating a 2465 kHz signal which is easily rejected by the IF amplifier. Difference product 1465 kHz – 1000 kHz creates a signal of 465 kHz. The mixer output of 465 kHz is easily amplified by the IF amplifier. The superheterodyne system allows the selectivity of the radio receiver to be achieved in the IF amplifier.

Receiving single side band

The superheterodyne receiver will not receive SSB signals without modification. As the SSB signal has no constant carrier, the rectified SSB signal is unintelligible. We need to reinsert the carrier into the signal path, just before the rectification. Figure 3.21 shows a modified superheterodyne receiver which has a beat frequency oscillator (BFO) or carrier insertion oscillator (COO).

The BFO 'recreates' the missing carrier within the output of the IF amplifier, immediately prior to rectification. The signal now presented to the diode looks like a correctly shaped AM waveform, which when rectified sounds intelligible. The frequency of the LO and the BFO needs to be very accurate with respect to the transmitted frequency otherwise the audio recovered from the diode rectifier quickly becomes unintelligible. The effect of an incorrectly tuned SSB receiver can be compared to a 'Donald Duck' voice!

The BFO is useful in any system where the carrier has been removed, suppressed or is not constant and is required to be reinserted prior to diode rectification or detection.

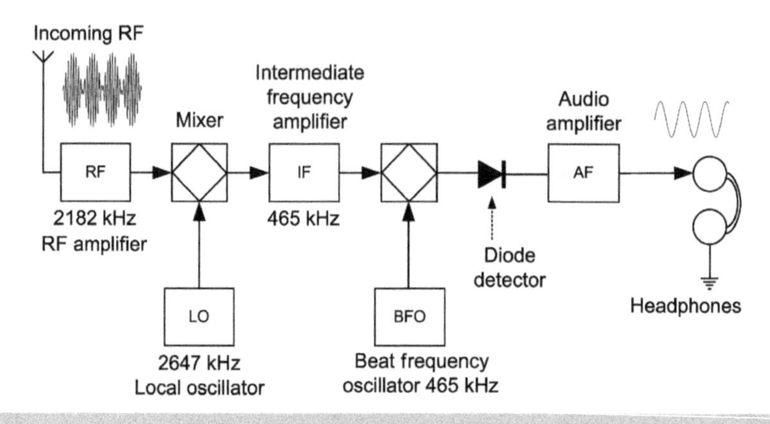

▲ **Figure 3.21** *SSB superheterodyne receiver*

Receiving frequency modulation

To receive an FM signal we need a signal detector that is sensitive to a change in frequency. In FM we usually call the detector a discriminator. A number of FM signal discriminators have been developed. Popular discriminators used two tuned circuits, slightly above and slightly below the IF. As the FM signal deviated above and below the central frequency, each of the two LC circuits would receive more or less signal. The output from these two coils is the audio signal. A more modern method is to use a PLL which allows the complete discriminator to be designed into one integrated circuit.

Figure 3.22 shows a traditional FM receiver with two coils for its discriminator. Some discriminators respond to changes in signal strength; to avoid this 'limiters' are used to increase and limit the size of the RF signal. As the limiter is amplifying the signal by a very large amount, when the transmitter ceases transmission, the discriminator detects random noise in the output of the IF/limiter causing a very loud hiss in the loud speaker. FM radio receivers have a 'squelch' control that can be adjusted to mute the audio output whenever the transmitter is silent to mute this hiss. Due to the characteristic of discriminators, FM receivers/discriminators will always listen/follow the strongest signal, ignoring the weaker signal completely. If the stronger signal is just slightly stronger, the FM receiver will 'capture' it ignoring other transmitters on the same frequency. This is the capture effect. Whereas AM receivers, output all signals received on a specified frequency.

Narrow band direct printing is a form of frequency modulation. A carrier frequency is frequency modulated by a digital information signal at 100 baud. The deviation is 170 Hz. A carrier for example could vary from 518000 to 518170 Hz, following the digital data from an information stream. These signals can be demodulated with a specialist

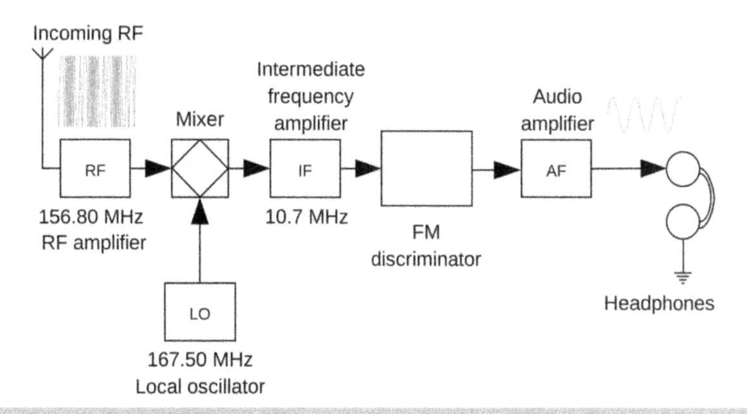

▲ **Figure 3.22** *FM superheterodyne receiver*

discriminator, typically a PLL. Intended uses include the transmission and reception of navigational warnings directly printed onto paper.

Receiver Characteristics

For radio receivers to be effective they need to reach certain minimum standards. These include:

- *Sensitivity*: The weakest signal we can receive, typically 0.2 μV for 1 W of audio output.
- *Dynamic range*: Can this receiver allow you to listen to a weak signal when a few kilohertz away is a local high power station? (Can you listen to a weak signal without the strong adjacent signal overloading our inputs?)
- *Signal-to-noise ratio*: It is the ratio of the signal strength to noise level, gives an indication of how noisy our receiver is and the minimum level of signal we need.
- *Image channel rejection*: How well do we reject signals from the mixer that are exactly 'one local oscillator frequency' away from the required carrier frequency?
- *Adjacent channel rejection*: How well do we ignore a station on the adjacent channel, perhaps just a few kilohertz away?
- *Bandwidth*: How wide is our receiver channel, 5 or 9 kHz perhaps?

Digital modulation and demodulation techniques are now common on higher frequency bands. Digital techniques allow a higher data rate of signal transmission (we can send more data between the transmitter and receiver at a faster rate). Digital transmission is outside the scope of the book.

Global Maritime Distress and Safety System

Global Maritime Distress and Safety System (GMDSS) is an integrated distress and safety radio communications system that vessels above certain sizes (>300 GRT) are legally obliged to follow. It allows for the timely transmission and reception of messages relating to:

transmitting ship-to-shore distress alerts by at least two separate and independent means;

receiving shore-to-ship distress alerts;

transmitting and receiving ship-to-ship distress alerts;

transmitting and receiving search and rescue co-ordinating communications;

transmitting and receiving on-scene communications;

transmitting and (as required) receiving signals for locating;

transmitting and receiving maritime safety information;

transmitting and receiving general radio communications to and from shore-based radio systems or networks and

transmitting and receiving bridge-to-bridge communications.

In the GMDSS, the scheme covers areas based upon the vessels range from the coast:

Area A1: Within range of VHF coast stations which have digital selective calling (DSC) (about 20–30 miles).

Area A2: Beyond area A1, but within range of MF coastal stations which have DSC (about 100–150 miles).

Area A3: Beyond the areas of A1 and A2, but within coverage of geo-stationary maritime communication satellites. This covers the area between approximately 70°N and 70°S.

Area A4: The remaining sea areas; satellites positioned above the equator, cannot 'see' ships above 70°N or below 70°S.

A typical GMDSS installation consists of:

VHF transceiver with DSC to cover a distance of up to 30 miles (A1) (Figure 3.23);

▲ **Figure 3.23** *SAILOR 6222 DSC VHF Tx/RX.* © *Thrane & Thrane*

MF transceiver with DSC to cover distances up to 100 miles (A2) (Figure 3.24);

▲ **Figure 3.24** *SAILOR 6301 DSC MF/HF Tx/RX.* © *Thrane & Thrane*

Satellite terminal using the INMARSAT service to cover the A3 region (Figure 3.25);

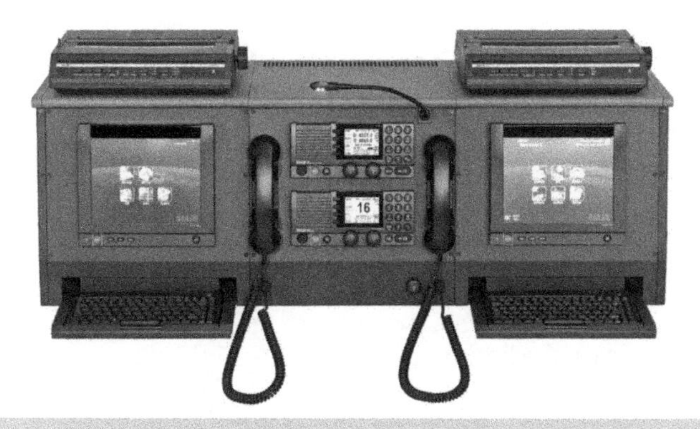

▲ **Figure 3.25** *SAILOR Satellite plus DSC VHF/MF/HF Tx/Rx.* © *Thrane & Thrane*

HF transceiver with DSC to cover the A3/A4 (Figure 3.26).

▲ **Figure 3.26** *SAILOR 6301 DSC MF/HF Tx/RX. © Thrane & Thrane*

Cost effective GMDSS installations combine the above into compact consoles containing a typical fit for a vessel sailing in unrestricted areas (A1/A2/A3/A4) (Figure 3.27):

- Two DSC VHF;
- Two DSC MF/HF 150 W, 250 W or 500 W;
- One mini-C GMDSS satellite system.

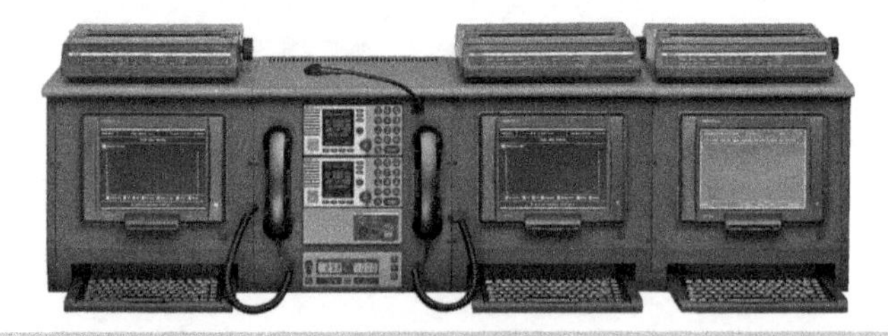

▲ **Figure 3.27** *SAILOR A4 Console Satellite DSC MF/HF Tx/RX. © Thrane & Thrane*

Vessels trading within A1/A2/A3 regions could have an integrated system containing:

- DSC-VHF/FM Radiotelephone,
- MF/HF Radiotelephone/DSC Watch Receiver/Telex,
- Standard-C satellite system.

Vessels trading within A1/A2 regions could have an integrated system containing:

- Two DSC VHF (one on the bridge),
- One DSC MF/HF 150 W, 250 W or 500 W (Figure 3.28).

▲ **Figure 3.28** *SAILOR A2 Console DSC MF/HF Tx/RX.* © *Thrane & Thrane*

Vessels trading within the A1 region could have an integrated system containing:

- Two DSC VHF

All GMDSS vessels, irrespective of trading area, require the following:

- One 406 MHz EPIRB (Emergency Position Indicating Radio Beacon) (Figure 3.29)
- One SART (Search and Rescue transponder) (Figure 3.30)
- One NAVTEX receiver (Figure 3.31)
- Three Portable GMDSS VHF (Figure 3.32)

▲ **Figure 3.29** *SAILOR EPIRB (Emergency Position Indicating Radio Beacon). © Thrane & Thrane*

The rules for maintenance of GMDSS installations are complex but can be reduced to the following; to ensure ability to send and receive distress alerts the following choices are available:

- use shore maintenance services;
- carry a qualified person, with spares and manuals, who can maintain the installation or
- duplicate the installation.

Some administrations force ship owners to use one option only.

The most common method to ensure the availability of the equipment is by duplication.

▲ **Figure 3.30** *SAILOR SART (Search and Rescue transponder).* © *Thrane & Thrane*

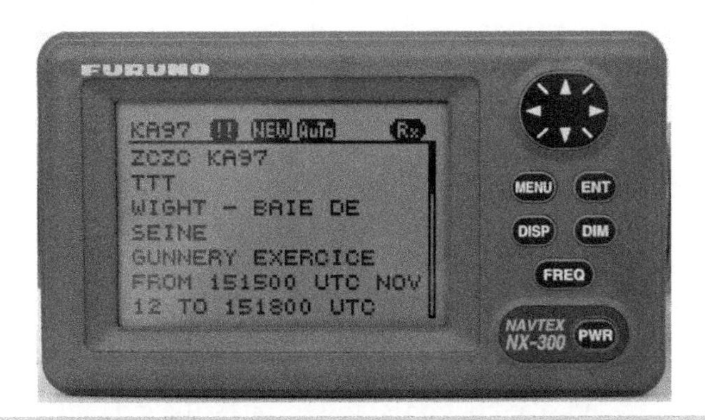

▲ **Figure 3.31** *Furuno NAVTEX*

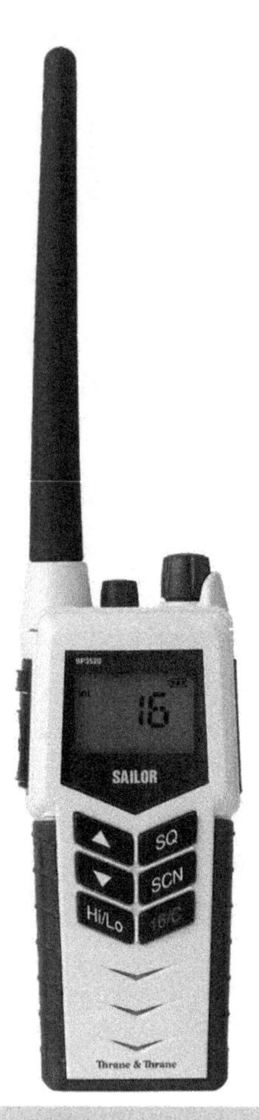

▲ **Figure 3.32** *SAILOR SP3520 Portable VHF GMDSS. © Thrane & Thrane*

It is now common to see GMDSS installations that consist of two DSC VHF, two DSC MF/HF and two satellite systems. This combined with Navtex, EPRIB, SART and portable VHF gives worldwide reliable coverage and availability.

Digital selective calling (DSC) enables the automation of many distress functions. A DSC transmitter sends a series of digital characters to initiate a call. If a vessel is in distress, when triggered, the DSC controller will set transmitters to the correct frequency, tune the antenna, then emit an internationally agreed message indicating the position and other details to enable a speedy rescue.

To trigger a Distress Alert, a 'red' button is pressed for 2 s. The 'red' button will be covered by a flap which should reduce the number of accidental false alerts. This button will initiate a DSC distress message sent by all available GMDSS connected systems. This will include VHF, MF/HF and satellite. The message will contain the vessel's MMSI number, its GPS position, time of position and nature of emergency, if the Master had time to select from the following list prior to pressing the initiate button. The non-selected danger is UNDESIGNATED DISTRESS, other specific dangers include fire/explosion, flooding, collision, grounding, listing and in danger of capsizing, sinking, disabled and adrift, or abandoning ship.

The DSC distress alert will be repeatedly transmitted either on all distress frequencies sequentially or as chosen by the operator, this is equipment specific. The DSC will stop transmission if the distress is acknowledged by a shore station or cancelled by the operator. The DSC can also be used for urgency and safety messages along with instigating calls to coast radio stations to set up telephone calls.

It is of the utmost importance that all regular servicing and testing of a GMDSS system is conducted thoroughly. Any self-test failures indicated by the GMDSS equipment should be logged and attended to without delay.

All equipment that is designed for unattended use is computer controlled with built-in test programs. Regular exercise of built-in test features, which are manufacturer specific, will highlight any shortcomings in the system and should be addressed without delay.

The DSC obtains its position information from an NMEA output of the GPS receiver. If this input fails the GMDSS system will issue a warning.

Batteries

For GMDSS and radio communications to work we need reliable power. To provide power to the GMDSS system during a blackout, we need a set of batteries that last the required minimum time. On a ship with an emergency generator the time is one hour. Without an emergency generator the batteries are required to last for six hours.

Two main types of batteries have been traditionally used for this function: Lead Acid and Nickel Cadmium. Other battery technology may be used if approved. The batteries require a daily, monthly and annual checks. The daily check is simply measuring the voltage drop when a load is placed on the battery. Monthly checks include checking the physical condition of the battery. Annual checks include those undertaken in port where the battery is discharged within the one hour or six hour limit, then recharged within the ten hour limit. As with all GMDSS testing, everything is recorded in the radio log book. Batteries within portable VHF handsets and EPIRBs are tested once per month during their regular checks.

4

SAMPLE QUESTIONS AND ANSWERS

1. A battery consists of 6 lead acid cells in series, each having an e.m.f. of 1.85V when discharged. The internal resistance of each cell is 0.1Ω.
   a) If a D.C. charging source of 120V is available, calculate the series resistance required at the start of the charge to set a charging current of 5A. State any assumptions made. (6 marks)
   b) Sketch the charging circuit showing the recommended position for the voltmeter and ammeter. (4 marks)
2. A lead acid battery measures 13.5V off load. A load of 10A was connected to the battery. The on load voltage was measured as 13.25V.
   a) What is the battery internal resistance? (5 marks)
   b) If the load was left connected and the terminal voltage dropped 0.25V in the first hour, estimate how long the battery would last until the terminal voltage dropped to 12.5V. State any assumptions made. (5 marks)
3. You have a 24V lead acid battery and a 100V DC voltage source.
   a) Sketch a circuit to connect the battery to the DC source and charge at approximately 4A. Include a voltmeter and ammeter. State any assumptions made. (10 marks)

4.

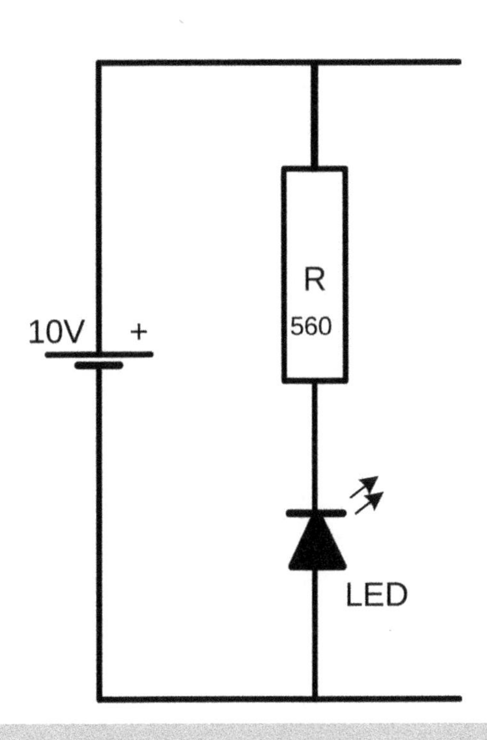

▲ **Figure 4.1**

a) In Figure 4.1 calculate the current flow through the resistor. (2 marks)

b) State any assumptions and justifications. (2 marks)

5.

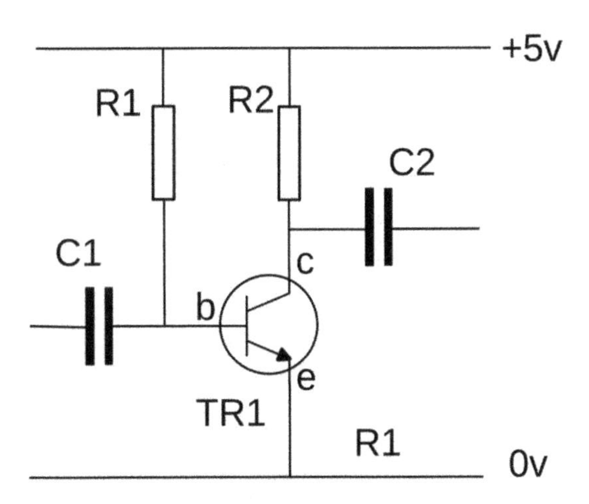

▲ **Figure 4.2**

a) In Figure 4.2 calculate the voltages on each transistor electrode. State any assumptions and justifications. (4 marks)

b) Calculate the current through R1. (1 mark)

c) Sketch the input and output waveforms when a sine wave is applied to the left of the input capacitor C1, measured at the output of C2. Mark your sketch with voltage and phase. (5 marks)

6. a) Calculate the power dissipated by a 1kΩ resistor when 240VAC is applied. State any assumptions and justifications. (4 marks)

   b) What will the power dissipation be if the voltage is changed from AC to DC?

   State any assumptions and justifications. (4 marks)

   c) If the resistor had a 10 percent tolerance, what would be the minimum and maximum dissipation? (2 marks)

7. a) Using Figure 4.3, calculate the circuit resistance using both methods discussed. (4 marks)

   b) What is the current flowing through R3? (2 marks)

   c) What is the power dissipated by the complete circuit? (2 marks)

   d) What is the voltage across R5? (2 marks)

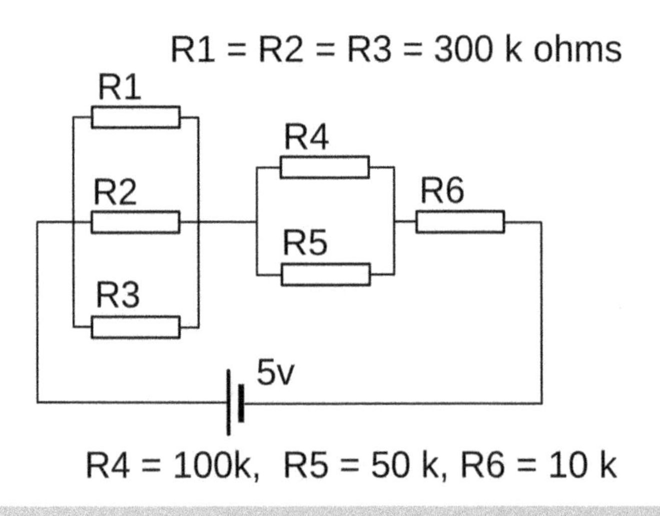

▲ Figure 4.3

8.

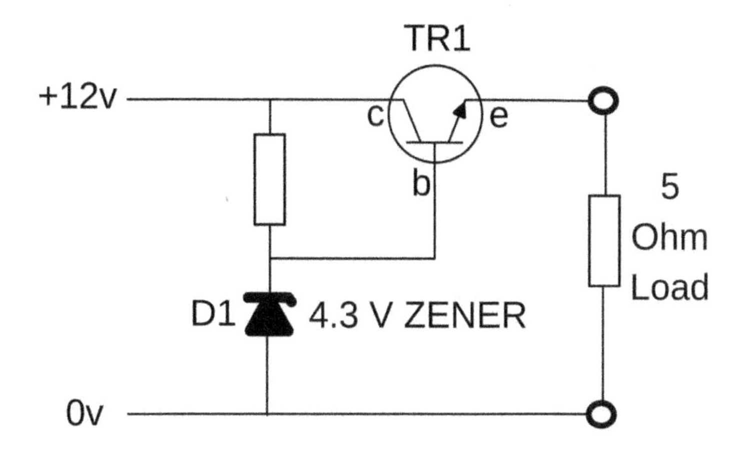

▲ Figure 4.4

a) Calculate the voltages on each transistor electrode. State assumptions. (2 marks)

b) Calculate the current flowing through the load resistor. (2 marks)

c) Calculate the voltage across the load resistor. (2 marks)

d) Calculate the voltage across the emitter–collector junction. (2 marks)

e) Calculate the power dissipated in the load resistor. (2 marks)

f) Calculate the power dissipated in the transistor. (2 marks)

9. A coil with a resistance 50Ω and inductance 0.50H is connected in series with a capacitance of 5μF across 240V, 50Hz AC supply.

a) Calculate the circuit impedance. (5 marks)

b) Calculate the circuit current. (2 marks)

c) Calculate the frequency at which the circuit's impedance is a minimum. (3 marks)

10. a) What are the key parts of a single board computer system? (5 marks)

b) Sketch a simple labelled single board computer system. (5 marks)

11. Describe the key characteristics of an NMEA 0183 system as fitted to large merchant vessels. (10 marks)

12. List reasons why electronic equipment fails. (10 marks)

13. Sketch and label a Radar transceiver. (10 marks)

14. Describe how a silicon transistor can amplify signals. (10 marks)

15. Why would you want to limit the amplitude of signals in radio transmitters? (10 marks)

Answers to Questions

1. a) Battery EMF = 1.85V x 6 = 11.1V

   Battery Internal Resistance = 0.1Ω x 6 = 0.6Ω

   Voltage dropped by IR = 0.6Ω x 5A = 3V

   Battery terminal voltage at 5A charge = 3V + 11.1V = 14.1V

   Voltage across series resistance = 120V – 14.1V = 105.9V

   Resistance needed $= \frac{105.9V}{5} = \underline{21.18\Omega}$

   Assumptions: that the voltage of 14.1V across the battery will charge it a 5A. Charging rate will decrease as the battery voltage rises with increasing charge.

   b)

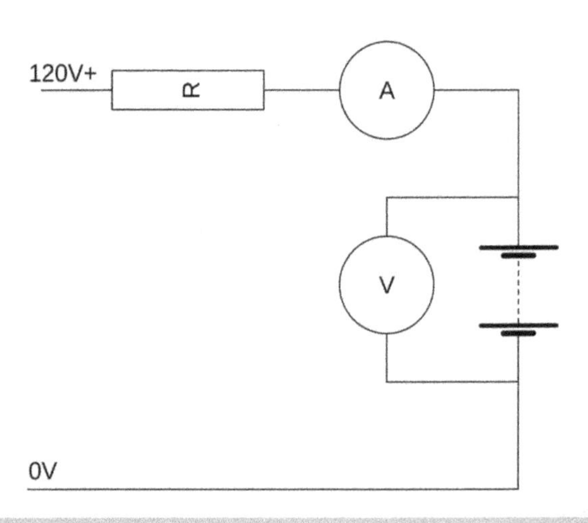

▲ Figure 4.5

2. a) Off load voltage = 13.5V, on load voltage = 13.25V

   Voltage drop due to internal resistance = 13.5V – 13.25V = 0.25V

   At 10A load, internal resistance $= \frac{0.25V}{10A} = \underline{0.025\Omega}$

   b) From 13.25V to 12.5V would take 3 hours.

   IE 13.25V

13.00V	1 hour
12.75V	2 hours
12.50V	3 hours

3. a)

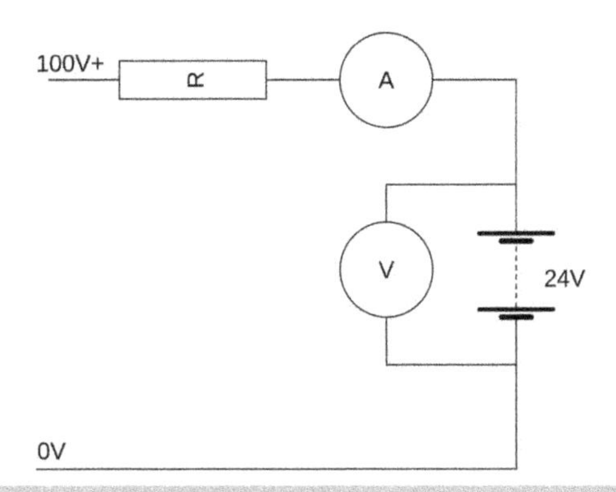

▲ Figure 4.6

R would equal 3.125 Ohms. This assumes a negligible internal battery resistance. From: 100V – 25V = 75V. R = V/I = 75/4 = 3.125 Ohms.

4. a) 10V – 2.1V = 7.9V across the resistor.

$$\mathbf{I} = \frac{7.9V}{560\Omega} = \underline{14.1mA}$$

   b) Assuming a small LED will have 2.1V across it when lit.

5. a) E = 0V due to being connected direct to 0V.

   B = 0.7V due to being forward biased via R1 and the transistor being silicon.

   C = 2.5V approximately due to this being an AC amplifier. Assuming it is linear in operation 2.5V would be ideal as its midrange. The value of R1 would control the voltage at C.

   b) Assuming the gain of the transistor is 100, the collector current = 100mA, the base current will be 1mA.

   R1 current = 1mA.

   c)

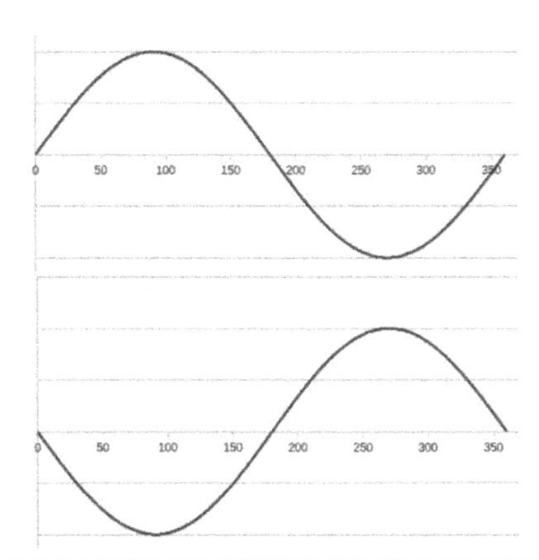

▲ Figure 4.7

In Figure 4.7 the input and output waveforms are shown 180 degrees out of phase. The relative size of each signal depends upon the gain of the transistor and value of the collector resistor.

6. a) $P = \frac{V^2}{R} = \frac{240^2}{1000} = \underline{57.6W}$

   b) All AC voltages, currents and powers are specified in RMS values unless otherwise specified. Therefore the power dissipated will be the same in both cases.

   c) $P = \frac{V^2}{R} = \frac{240^2}{1100} = \underline{52.4W}$

   $P = \frac{V^2}{R} = \frac{240^2}{900} = \underline{64W}$

7. a) R1 & R2 & R3 all equal 100k so three 100k in parallel equals:

   $\frac{100,000}{3} = 33.3\text{k or}$

   $1/100k + 1/100k + 1/100 = 1/Rt(100k) = 33.3k$

   $R4//R5 = \frac{100k \times 50k}{100k + 50k} = 33.3k$ *or*

   $R4//R5 = 1/100k + 1/50k = 1/Rt(50k) = 33.3k$

   Plus the remaining R6 of 10k $= 33.3k + 33.3k + 10 = \underline{76.6k\Omega}$

   b) Total circuit current $I_T = V/R = \frac{5V}{76.6k} = 65.3uA$ or

   $I(R3)$ = one third of the circuit current therefore $\frac{65.3uA}{3} = \underline{21.8uA}$

   c) $P = VI = 5V \times 65.3uA = \underline{326.5uW}$

   d) First we calculate the voltage across R4//R5

   $V = IR = 65.3uA \times 33.6k = 2.19V$

   V across R5 = 2.19V

8. a) TR1 C = 12V (direct connection to 12V+)

   TR1 B = 4.3V (due to 4.3V developed across the zener diode D1)

   TR1 E = 5V (approximate value due to TR1 being forward biased, conducting therefore 0.7V is developed across the emitter-base junction. 4.3V + 0.7V = 5V)

   b) $I = V/R = 5V/5\Omega = \underline{1A}$

   c) It must be 5V due to its connection to TR1 Emitter $\underline{5V}$

   d) $12V - 5V = \underline{7V}$

   e) $P = VI = 5V \times 5A = \underline{25W}$

   f) $P = VI = 7V \times 5A = \underline{35W}$

9. a) XL = 2πfL XC = 1/(2πfC)

XL = 2 x π x 50 x 0.5 = 157.1Ω

XC = 1 x (2 x Ω x 50 x 5x10-⁶) = 636.6Ω

$$Z = \sqrt{R^2 + (X_L - X_C)^2}$$

$$Z = \sqrt{25^2 + (157.1 - 636.6)^2}$$

$\underline{Z = 479.6\Omega}$

b) I = V/R = 240V / 479.6Ω = $\underline{0.5A}$

c) $$F_{osc} = \frac{1}{2\pi\sqrt{LC}}$$

$$= F_{osc} = \frac{1}{2\pi\sqrt{0.005 \times 5E10^{-6}}} = \underline{1007Hz}$$

10. a) Central Processor Unit (CPU)

Read Only Memory (ROM)

Random Access Memory (RAM)

Digital Input/Output (IO)

Serial Input/Output (SIO)

Analog Input/Output (AIO)

Network Input/Output (NIO)

b)

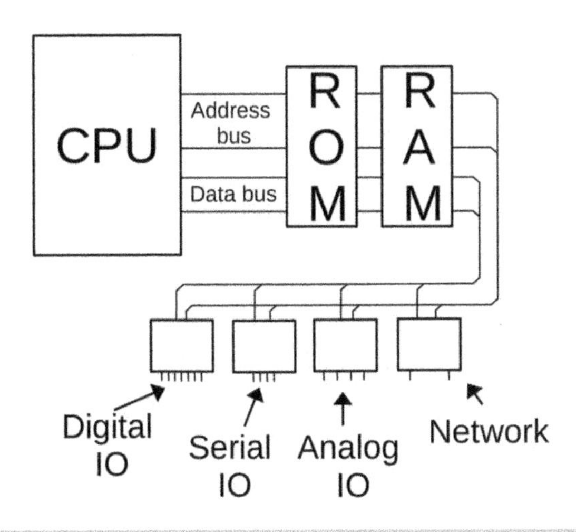

▲ Figure 4.8

11. NEMA 0183 systems are used to transmit and receive data between navigation equipment on large merchant vessels.

    It is a serial based system using two wires to communicate from one equipment to another. It does not daisy chain to multiple equipments but could be. The NEMA output and input of each equipment is isolated from ground and each equipment. If any part of the navigational equipment is catastrophically destroyed, the isolation provided by the NMEA interface isolates each equipment from all the others.

    It is a slow data transfer system, using plain text, but this means it is easy to diagnose faults when one occurs. You can observe data being transmitted or received with an oscilloscope as logic levels or you can connect a PC serial port and read the actual text information.

12. Heat (electronic components do not like heat, it causes early failures).

    Vibration (causes solders joints and pug/sockets to become open circuit).

    Moisture (excess moisture can seep into components, changing their characteristics).

    Static (electrostatic shock caused during manufacture or maintenance can destroy components or impair them).

    Strain and stress (can cause connections to break).

13.

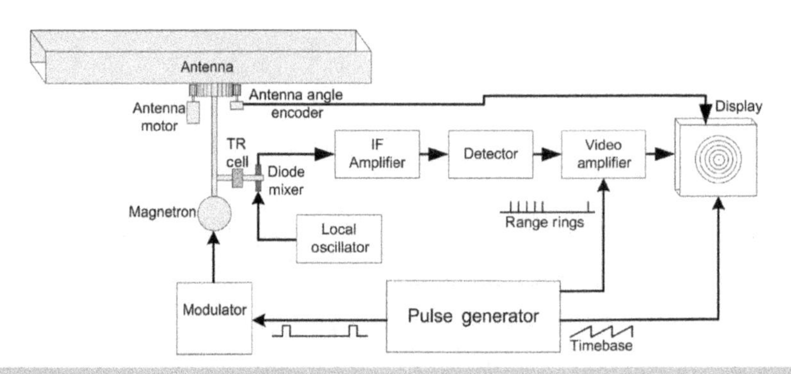

▲ **Figure 4.9** *shows a rather complete sketch of a radar system.*

You need to be able to sketch, in the correct positions, the following:

Magnetron

TR Cell

Antenna

Modulator

Local oscillator

Diode mixer

IF

Detector

Display

Pulse generator

14. Transistors amplify in the following way. For an NPN transistor, via a resistor, a voltage is fed to the transistor collector terminal. A very small current flows from the emitter to the collector. If a small current is injected into the base, the small current causes a much larger current to flow from the emitter to the collector.

    In more detail, the base–emitter junction is weakly forward biased. The current flowing into the base, causes the base-emitter junction to be more forward biased. This causes much more current to flow from the emitter to the collector. A small change in base current causes a larger change in collector current. This is current amplification. If we used electron current flow, base current flows out of the base.

15. If you have a correctly adjusted radio transmitter that transmits speech well, the listener will understand your voice! If you shout into the microphone, without limiting, you will overload various stages within the transmitter.

    With an SSB transmitter you will get distortion, splatter (various noise artefacts appearing on adjacent channels).

    With an FM transmitter, when you over modulate, your carrier frequency sweeps adjacent channels, preventing communication on them and spends less time in your channel, so your listener will not be able to understand you.

    In summary, we limit the amplitude of the voice signal immediately on input to the transmitter to prevent distortion and unwanted out-of-channel interference with other users.

INDEX

REEDS MARINE ENGINEERING AND TECHNOLOGY SERIES

Vol. 1 Mathematics for Marine Engineers
Kevin Corner, Leslie Jackson and William Embleton
ISBN 9781408175552

Vol. 2 Applied Mechanics for Marine Engineers
Leslie Jackson and William Embleton
ISBN 9780713672589

Vol. 3 Applied Heat for Marine Engineers
Leslie Jackson and William Embleton
ISBN 9780713667332

Vol. 4 Naval Architecture for Marine Engineers
E A Stokoe
ISBN 9780713667349

Vol. 5 Ship Construction for Marine Students
E A Stokoe
ISBN 9780713671780

Vol. 6 Basic Electrotechnology for Marine Engineers
Christopher Lavers, Edmund G R Kraal and Stanley Buyers
ISBN 9781408176061

Vol. 7 Advanced Electrotechnology for Marine Engineers
Edmund G R Kraal
ISBN 9780713676846

Vol. 8 General Engineering Knowledge for Marine Engineers
Paul A Russell, Leslie Jackson and Thomas D Morton
ISBN 9781408175965

Vol. 9 Steam Engineering Knowledge for Marine Engineers
Thomas D Morton
ISBN 9780713667363

Vol. 10 Instrumentation and Control Systems
Gordon Boyd and Leslie Jackson
ISBN 9781408175590

Vol. 11 Engineering Drawings for Marine Engineers
H G Beck
ISBN 9780713678574

Vol. 12 Motor Engineering Knowledge for Marine Engineers
Paul A Russell, Leslie Jackson and Thomas D Morton
ISBN 9781408175996

Vol. 13 Ship Stability, Resistance and Powering
Christopher Patterson and Jonathan Ridley
ISBN 9781408176122

Vol. 14 Stealth Warship Technology
Christopher Lavers
ISBN 9781408175255

Vol. 15 Electronics, Navigational Aids and Radio Theory for Electrotechnical Officers
Steve Richards
ISBN 9781408176092

Vol. 16 Electrical Power Systems for Marine Engineers
Gordon Boyd and Fred Taylor
ISBN 9781472968463